Computational Music Science

About this series - The CMS series covers all topics dealing with essential usage of mathematics for the formal conceptualization, modeling, theory, computation, and technology in music. The series publishes peer-reviewed only works.

Comprehensiveness - The series comprises symbolic, physical, and psychological reality, including areas such as mathematical music theory, musical acoustics, performance theory, sound engineering, music information retrieval, AI in music, programming, soft- and hardware for musical analysis, composition, performance, and gesture. The CMS series also includes mathematically oriented or computational aspects of music semiotics, philosophy, and psychology.

Quality - All volumes in the CMS series are published according to rigorous peer review, based on the editors' preview and selection and adequate refereeing by independent experts.

Collaboration - The editors of this series act in strong collaboration with the Society for Mathematics and Computation in Music and other professional societies and institutions.

Should an author wish to submit a manuscript, please note that this can be done by directly contacting the series Editorial Board, which is in charge of the peer-review process.

More information about this series at https://link.springer.com/bookseries/8349

Guerino Mazzola • Sangeeta Dey
Zilu Chen • Yan Pang

Functorial Semiotics for Creativity in Music and Mathematics

Springer

Guerino Mazzola
School of Music
University of Minnesota
Minneapolis, MN, USA

Sangeeta Dey
School of Music
University of Minnesota
Minneapolis, MN, USA

Zilu Chen
New York University
New York, NY, USA

Yan Pang
School of Music
Yan Pang Create, LLC
Minneapolis, MN, USA

ISSN 1868-0305 ISSN 1868-0313 (electronic)
Computational Music Science
ISBN 978-3-030-85189-7 ISBN 978-3-030-85190-3 (eBook)
https://doi.org/10.1007/978-3-030-85190-3

This Springer imprint is published by the registered company Springer Nature Switzerland AG
The registered company address is: Gewerbestrasse 11, 6330 Cham, Switzerland

Dedicated to Ida Mazzola,
Guerino's wife and center of hope.

Preface

Most of Guerino Mazzola'a books deal with the application of mathematics to music in theory, creativity, performance, composition, and embodiment. This implicitly implies that mathematics is meaningful for understanding music. This role of mathematics could be compared to its role for physics, reaching from the formal or formulaic description to the modeling and conceptualization of music-related facts, even in the limit claiming that the essence of the field (physics or music) is mathematical in its ontology, such as forwarded by Roger Penrose or Louis Michel for physics and Pythagoras or Wilhelm Leibniz for music, among others.

This conceptual direction has one remarkable deficiency: it emphasizes the pure form of the mathematical approach, in other words, it fosters abstraction against semiotic aspects of these phenomena. In physics, this might be less relevant, physics works perfectly without dealing with the semantics of elementary particles. But music is different, and especially its creative perspective. As was developed in Mazzola's book on *Musical Creativity* [32], creativity is always the process of extending a given semiotic system—which one also calls a "semiotic"—by adding a set of new signs to the given semiotic.

Understanding music therefore includes understanding its creative kernel, music is not a given world like physics, its substance is in constant expansion, even with regard to the very definition of the concept of music. The mathematical conceptualization of music therefore is not an ideal task that would satisfy physics, we also have to investigate the creative enrichment in musical developments to touch the fundamental nature of this art, science, and technology.

This was the reason why Mazzola published a paper *Functorial Semiotics for Creativity* [35] that deals with a mathematical theory of semiotics. His intention was to start a valid mathematical description of the classical semiotic approaches from their creators such as Charles Sanders Peirce, Ferdinand de Saussure, Louis Hjelmslev, and Roland Barthes.

The present book is an extension of that paper and does not only deal with musical creatvity, but more generally with a mathematical foundation of semiotics and its role for human creativity in general.

The role of mathematical paradigms should be mentioned here. As has become a standard point of view in post-Grothendieckian mathematics, the Yoneda Lemma also plays a crucial role in mathematical music theory and more precisely in its topos theory. For our approach it is however essential to take the Yoneda Lemma as a semiotic statement in disguise. The abstract object X in a category is replaced by its presheaf $Hom(?, X)$, which we denote by $@X$. This means that understanding X is accessed by understanding all the morphism perspectives $f : Y \to X$ in $@X$, i.e., the 'meaning' of X—its isomorphism class, if you mind—is explicated by the collection of its perspectives when viewed from all possible domains Y in the given category.

Our approach will inherit this "Yoneda philosophy" [24] that an object gains its 'meaning' or 'content' via the 'integral' of all its perspectives $f : Y \to X$. This semantic reinterpretation of the Yoneda Lemma is in fact the starting point of the "semantic topology", which we introduce in this book, a topology that works for any category, but is motivated by the Yoneda philosophy and the semiotic categories to be introduced here. We believe that the semiotic interpretation of Yoneda's Lemma and our subsequent extensions are an enrichment of the mathematical formalism that could connect its "empty form" to what has "content" in art and science.

Essentially, this book thematizes three steps of scientific progress: (1) *the semiotic nature of creativity,* (2) *the mathematical shaping of semiotics for its operationalization, and* (3) *the necessity to use* (2) *for a future computational creativity in AI.*

The authors are pleased to acknowledge the strong support for writing such a treatise from Springer's science editors Francesca Bonadei and Thomas Hempfling.

Minneapolis, June 2021

Guerino Mazzola, Yan Pang
Zilu Chen, Sangeeta Dey

Contents

Part IV Applications and Consequences

Part V Conclusions

Part VI References, Index

Part I

Orientation

1

Motivation and Background

Summary. We present the motivation for the development of this *functorial semiotics* as a bridge between Human Intelligence and Artificial Intelligence.

– Σ –

In recent computational creativity research [44], the question of whether machines or algorithms can be creative was raised in a fairy algorithmic spirit, meaning that machines might be creative in the framework of a rich combinatorial performance. It is however crucial to understand that creativity by its very nature has to add new signs, and to extend a given semiotic system. This is made evident in a thorough discussion of the process of musical creativity [32]. In other words, understanding creativity and the option of giving it a computational drive heavily depend on the question of how new signs, new contents, could be conceived in the framework of a computational formalism. For this reason we have investigated the question concerning a formal, or rather: mathematically precise representation of semiotics. But we have not (yet) applied this investigation to an equally mathematical understanding of the creative process as such. This book presents probably the first, and unavoidable step towards such a theory.

Semiotics has been developed following the classical Latin definition of a sign: "aliquid pro aliquo". It defines a sign as a threefold entity that comprises its expression (aliquid), its content (aliquo), and the transitional step, its signification (pro). This simple conceptual architecture is enriched by the syntactical combinatorics and by the pragmatics of sign usage. Ferdinand de Saussure [39] has added six dichotomies in his semiology, and Louis Hjelmslev [15] has enriched the semiotic system by two "vertical" extensions: *connotation* for expressive entities that are proper signs, and *metatheory* for contents that are proper signs; we could add a third in Hjelmslev's spirit, and also conceived in Charles Sanders Peirce's ternary sign architecture: representamen, interpretant, and object: *motivation* for significations that are proper signs.

This setup is however not a formal or mathematical framework. Many details are arcane, above all the signification process. And when we are dealing

G. Mazzola et al., *Functorial Semiotics for Creativity in Music and Mathematics*, Computational Music Science, https://doi.org/10.1007/978-3-030-85190-3_1

with the question of semantics, there is a dramatic lack of understanding what the meaning of a sign comprises. If, for example, you are dealing with the concept of a prime number, its meaning is given by the definition: a prime number is a natural number > 1 that has as divisors only the number one and the number itself. But this straightforward information does not cover what the concept means in its overall signification. For example, does this also mean that every even number > 2 is the sum of two prime numbers (Goldbach's conjecture)? Does it mean that there are infinitely many prime numbers?, etc. The content of the concept of a prime number covers much more than its original definition. Or, if we deal with numbers, their mathematical definition is not sufficient to understand their meaning. This could also include contents that are not mathematical, but, for example, historical, mystical, religious, etc. A comprehensive semiotic theory must also include such diverse contents. Otherwise, the process of signification remains fragmentary.

1.1 Local and Global Contents

This means that the meaning of a sign is not only its 'local' content, but comprises a variety of meanings that can be retrieved from a broader 'global' signification context. Without the inclusion of such global semantic perspectives, semiotics cannot claim to be representative of what humans perceive and process when they deal with semiotic instances. In creative processes such global aspects are crucial since one might create new contents that have not been anticipated from the given 'local' information. We also refer to the field of *frame semantics*, as proposed by Charles J. Fillmore [12], where frames represent structures of global semantics. Global semantics has also been addressed in Joseph Goguen's proposal of an "Algebraic Semiotics" [13], where he discusses a formalized semiotics with context-sensitive semantics: "Although our official name for this approach is **algebraic semiotics**, it might also be called **structural semiotics** to emphasize that meaning is structural, or (in its philosophic guise) even **morphic semiotics**, to emphasize that meaning is dynamic, contextual, embodied and social." Our approach to semiotics is different from Goguen's in that our system of semiotics is a big comprising category, while his approach is focused on restricted algebraic semiotic systems, not categories, which are connected via specific morphisms. Our approach can easily be seen to absorb his architecture as a set of relatively small subsystems. Moreover, Goguen's approach does not make explicit the Hjelmslev perspective of connotation, metatheory, and signification.

In what follows, we shall propose a semiotic system that includes such global options. A basic idea of such an approach can be traced from the famous Yoneda Lemma in mathematical category theory. In category theory, an object X is a sign in the chosen category $\mathcal{C}$. Its identity is an abstract fact, and this could be understood as its content, its identification qua object in $\mathcal{C}$. But its signification in $\mathcal{C}$ is not understood by this abstract identification. Its role and meaning to the given category $\mathcal{C}$ is only understood when one considers its contravariant functor[1] $@X := Hom(-, X)$. The functor $@X$ of X determines its role within $\mathcal{C}$. The content of X in $\mathcal{C}$ is much more than the abstract identity of X. It comprises all its morphisms $Y \to X \in Y@X = Hom(Y, X)$ and their relationships under transitions $Z \to Y$ from "address" Y to "address" Z. This exemplifies the difference between 'local' content and 'global' content of X. In this sense, any sign in a comprehensive semiotics has a global content that by far transcends its direct local content.

1.2 The Artificial Intelligence Problem

We should also mention the Artificial Intelligence (AI) field in computer science, where a certain industrially motivated hype has been around, especially suggesting that AI is on its way to become equivalent if not superior to Human Intelligence (HI). In the AI development, there have been three stages: (1) expert systems, the simulation of the brain's activities, (2) connectionist systems, typically realized by artificial neural networks (ANNs), simulating the human neuronal networks, (3) embodied AI, including not only the brain and the nerves, but also the body and its inherent architecture. The historical development from (1) to (2) to (3) was always induced by deficiencies in the previous version.

At present, the dominant AI technology is based upon the connectionist paradigm, where ANNs are used to simulate intelligent behavior. For example, training an ANN by a presentation of 10,000 images of a cat's face, may induce that ANN to "recognize" the next cat image as being a cat. This wording is however misleading since the ANN does not understand anything about what is a cat, it simply performs a pattern matching operation without any semantic depth. ANNs are blind with respect to semiotic aspects, this might be best illustrated by Cybenko's theorem [8], which states that for any given input and output, one may construct an ANN that produces the given output from the given input.

[1] We stick to the categorical notations explained in [27] and used in all his previous publications relating to category theory. Let us recall this notational approach. For any two objects X, Y in a category $\mathcal{C}$, we denote $X@_{\mathcal{C}}Y$, or $X@Y$ if the underlying category is clear, for $Hom_{\mathcal{C}}(X, Y)$. Accordingly, $@X$ denotes the contravariant functor $\mathcal{C} \to \mathbf{Ens} : Y \mapsto Y@X$, while $X@$ denotes the corresponding covariant functor. The category of contravariant functors (presheaves) $\mathcal{C} \to \mathbf{Ens}$ is denoted by $\mathcal{C}^{@}$. For such a presheaf F, we denote by $X@F$ its value at object X. This is justified by the Yoneda Lemma: In the category $\mathcal{C}^{@}$, there is a canonical bijection $(@X)@F \overset{\sim}{\to} X@F$.

1.3 Mathematical Concept Constructions

The main effort in this book is dedicated to the construction of a mathematical architecture that takes care of the semiotic phenomenon. We shall introduce a type of *semiotic category* that explicitly models the 'social life' of semiotic components such as signs with their expressive, semantic, and significative coordinates. But such a construction is not sufficient, the analysis of a semiotic category requires its own mathematical tools and methods. This relates to the question of how mathematical understanding of semiotic phenomena, when modeled within a semiotic category, can be deepened. To put the question into more intuitive words, we shall investigate mathematical 'coordinate functions' to pinpoint the semiotic dynamics.

Such 'coordinate functions' will be modeled following generalizations of Yoneda's Lemma, and introducing a simple, but very efficient topology, the *semantic topology* on categories. This construction is motivated by the very definition of semiotic categories, but it works for any category. It will be used to discuss Čech cohomology that describes the local-global relationship of spaces of above-mentioned 'coordinate functions'.

1.3.1 Size of Categories

One general problem arises when one deals with general categorical architectures: the size of categories, i.e., the problem of whether a category is or is not small (set-like). This one is related to the foundational questions stemming from set theory, such as the set of all sets, non-existent in Z-F set theory, for example. We shall not discuss such aspects for two reasons: first, if the reader feels uncomfortable with our discourse, he/she may still strictly filter our exposé to small categories; second, we do not agree with all of the Z-F principles, and follow Paul Finsler's approach to set theory [10], and not any "paper mathematics" approach (oral citation from Finsler to the author), as described by William Lawvere [19], for example, but this is not what this book is about. And also observe that Lawvere's term "functorial semantics" [20] is not a semiotic term, it is a purely technical term that describes certain (co)adjoint functors. So, if the reader feels/perceives a crisis in such foundational perspectives, simply go down to small categories altogether, the conceptual architecture of this book is not about size of categories.

We shall denote by CAT the category of small categories. Viewing sets as categories with identities only (discrete categories), the category of sets **Ens** is a subcategory of CAT.

1.4 Some Ontology

Although semiotics is not focused on ontological issues, we shall have to deal with such topics because of the question about the semiotic 'traces' of ontological qualities. We therefore give a short summary of the ontological architecture that we developed with great success for music theory.

Ontology, includes three dimensions: realities, semiotics, and communication. It also includes the extension of ontology to the fourth dimension of embodiment. We call this extension "oniontology" for reasons that will become evident soon.

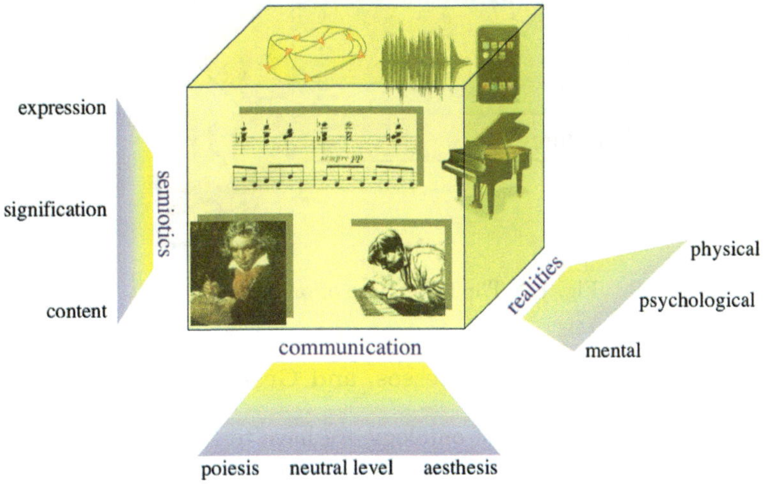

Fig. 1.1: The three-dimensional cube of ontology.

1.4.1 Ontology: Where, Why, and How

Ontology is the science of being. We are therefore discussing the ways of being that are shared by music and many other human activities. As shown in Figure 1.1, we view it as being as spanned by three 'dimensions', i.e., fundamental ways of being. The first one is the dimension of realities. Human reality has a threefold articulated reality: physics, psychology, and mentality. Mentality means that music has a symbolic reality, which it shares with mathematics. This answers the question of "where" something exists. This triple reality is quite similar to the three world model of Karl Popper and John Eccles [37].

The second dimension, semiotics, specifies that being is also one of meaningful expression. Human existence is also an expressive entity. This answers the question of "why" something is so important: it creates meaningful expressions, the signs that point to contents.

The third dimension, communication, stresses the fact that existence is also a shared being between a sender (for example the composer or musician), the message (typically the composition), and the receiver (the audience). Communication answers the question of "how" existence takes place.

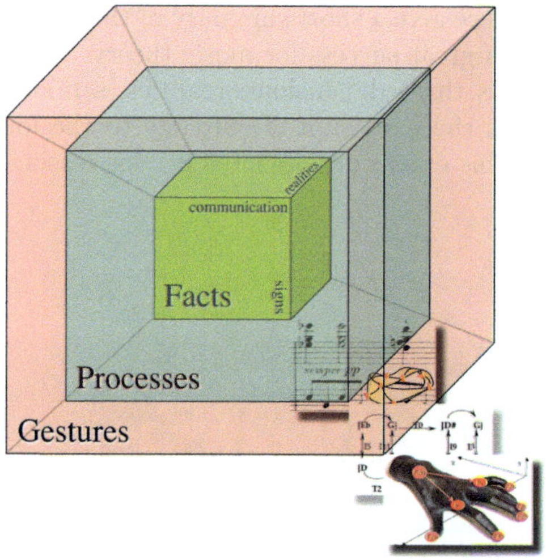

Fig. 1.2: The hypercube of oniontology.

1.4.2 Oniontology: Facts, Processes, and Gestures

Beyond the three dimensions of ontology, we have to be aware that existence is not only a being that is built from facts and finished results. It is strongly also processual, creative, and living in the very making of things. Human performance is a typical essence of an activity that lives, especially in the realm of 'improvisation', while being created. The fourth dimension, *embodiment*, deals with this aspect; it answers the question "how *to come into* being?" It is articulated in three values: facts, processes, and gestures. This fourth dimension of embodiment gives the cube of the three ontological dimensions a threefold aspect: ontology of facts, of processes, and of gestures. This four-dimensional display can be visualized as a threefold imbrication of the ontological cube, and this, as shown in Figure 1.2, turns out to be a threefold layering, similar to an onion. This is the reason why we coined this structure "oniontology"—it sounds funny, but it is an adequate terminology.

Part II

General Concepts

2

Semiotics

Summary. This chapter gives an overview of semiotics as developed by Charles Sanders Peirce, Ferdinand de Saussure, Louis Hjelmslev, and Roland Barthes.

– Σ –

2.1 Generalities about Signs

Summary. Semiotics is the dimension of meaning. It studies the structure of symbols and signs and their associated meanings. We will start by introducing the basic principles of semiotics: the different levels of symbols and the structure of musical symbolism. We will then discuss philosophers and linguists that studied semiotics, and explain their theories as they relate to music. We will apply the theories of Ferdinand de Saussure, who is perhaps the most influential scholar of semiotics, to analyze and exemplify music as a symbolic system. We will discuss the Babushka Principle of semiotics, which leads into such topics as connotation, motivation, and metatheory.

– Σ –

2.1.1 The Problematic Fundamental Position of Semiotics in Human Behavior

Summary. We give a short presentation of the seemingly mandatory/inevitable involvement of semiotic processing in humans.

– Σ –

Often, when one discusses semiotic issues, the statement is made that "everything has a meaning", that our fundamental mechanism of understanding what humans do, think, and feel, is to find the meaning of what is being presented. This is a widespread opinion which also shapes a number of basic scientific

G. Mazzola et al., *Functorial Semiotics for Creativity in Music and Mathematics*, Computational Music Science, https://doi.org/10.1007/978-3-030-85190-3_2

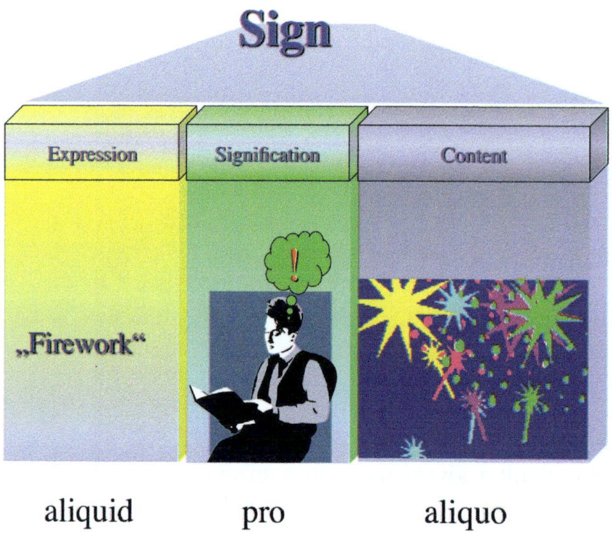

Fig. 2.1: The sign's anatomy.

approaches. Typically, gesture theory in its Anglo-American structure [21] is based upon the axiom that gestures are carriers of given contents, they don't generate contents, they only carry them from a sender to a receiver. The French approach to gestures, as developed by Paul Valéry, Charles Alunni, and others [28, Ch. 58], is however opposed to this perspective. They view gestures as being "wild" entities, which can be generators of semantic contents, and as such are presemiotic motors. Refer to Section 4.2 of this book for details concerning the presemiotic conditions of the semiotic architecture.

2.1.2 Definition of Signs

Signs are sets of symbols with a conventional meaning that is associated. For example, a musical score is a collection of symbols that are interpreted and translated into musical gestures and sound by professionally trained performers. A sign can be divided into three components: expression, signification and content. For example, when we read the word "firework", we read an expression, "firework", a set of characters, to which we associate a content (the sounding firework) via the signification process, see Figure 2.1.

2.2 De Saussure and Peirce: the Semiotic Architecture

Summary. Semiotics has been studied by linguists and philosophers for many years. The first attempts to define the components of a sign system were made in 1865 by the United States philosopher, Charles Sanders Peirce. After Peirce came Ferdinand de Saussure, Louis Hjelmslev, and Roland Barthes, each with differing views on the components of a sign system. In this chapter we will explore contributions made by the four aforementioned semiotic theorists and discuss the semiotic architecture that their theories illustrate.

– Σ –

2.2.1 Pierce

The American scholar Charles Sanders Peirce (see Figure 7.1) was one of the first theorists to explore the world of semiotics. Though he studied chemistry during his college education, Peirce was truly a Renaissance man, publishing research in semiotics, linguistics, psychology, philosophy, statistics, and economics among other fields. Peirce's philosophical analyses spanned from concepts of logic to theories about the mind. Semiotics was born from Peirce's desire for a system of analysis to describe the general use of signs, as inspired by their application in logic.

However, it is important to note that Peirce's contributions to the field of semiotics were in the form of what he called "semeiotic", which is fundamentally different than the field of semiotics that we know today. Peirce's studies focused on the existence of the "triad of representing relation," in which a symbol was comprised of an *object* (something being represented), a *representamen* (something different than the object that represents the object and is related to a third concept), and the *interpretant* (the concept that the representamen appeals to). However, this proposal was complex, lacking clarity as to where the representation began and the represented concept ended. Later semiotic scholars sought a more eloquent, if not simpler, model for symbolic representation.

Fig. 2.2: Ch. S. Peirce (1839-1914).

Fig. 2.3: F. de Saussure (1857-1913).

Fig. 2.4: L. Hjelmslev (1899-1965).

2.2.2 de Saussure

While Peirce laid the foundations for the field of semiotics, Ferdinand de Saussure is the true founding father of modern semiotics. Ferdinand de Saussure was a Swiss linguist and philosopher (see Figure 7.2). Saussure established himself as a brilliant linguist, when he published his first and only book *Memoir on the Original System of Vowels in the Indo-European Languages* as a student. Saussure was not a particularly prolific writer, but his few publications revolutionized the field of semiotics. Moreover, Saussure as instructor and professor had an immense influence in the field. His most influential work was a culmination of his lectures grafted together by his students and colleagues, published post-mortem in 1916. In this short introduction to linguistics entitled *Cours de Linguistique Générale* [39], Saussure proposed six dichotomies (to be discussed in Chapter 2.3) that describe the nature of linguistics or, more generally, semiotics.

In a structure reminiscent of Peirce's argument, Saussure suggested that symbols are comprised of three interacting components: the *signifier* (the symbol used to symbolize an object), the *signification* (the process of symbolization), and the *signified* (the object being symbolized). Though it has been modified by linguists such as Louis Hjelmslev and Roland Barthes, this theory forms the backbone of modern semiotics.

2.2.3 Hjelmslev

Louis Hjelmslev was a Danish linguist who, like Saussure and Peirce, was interested in semiotics (see Figure 7.3). Hjelmslev is best known for his theory of glossematics, which revised Saussure's version of semiotics and added new concepts. Of these new concepts, the most important was what is called "connotation", which will be described in Section 2.4. In addition to this, Hjelmslev revised the signifier/signification/signified structure that Saussure proposed. In his book *La stratification du langage* [15], Hjemslev instead proposed the structure of *expression* (the symbol), *relation* (the relationship between the symbol and the content, similar to the signification process), and *content* (the meaning expressed by the symbol).

While Saussure built the backbone of semiotics, Hjelmslev was responsible for much of its development and popularization. Hjelmslev's theories allowed semiotics to expand vastly, branching out to become useful in areas outside of linguistics and philosophy within which they had largely been contained. In fact, Hjelmslev's contribution of connotation explains the infinite expansion of sign systems that Peirce originally desired to capture with his object/representamen/interpretant model of semiotics.

For the purposes of this book, we will use a model of the semiotic architecture that combines the ideas of Hjelmslev and Saussure. The combined sign model is presented as expression, signification, and content. This model incorporates the ideas of expression and content from Hjelmslev's model, while maintaining signification from Saussure's model. We feel that this is the clearest and most intuitive explanation of the semiotic architecture for the English language. Given this model, we analyze music (and eventually expand out to more general areas), a vastly different topic than the traditional use of semiotics as a linguistic tool. Today, it is common to use semiotics to study sign systems in a wide variety of areas, but this was not always the case.

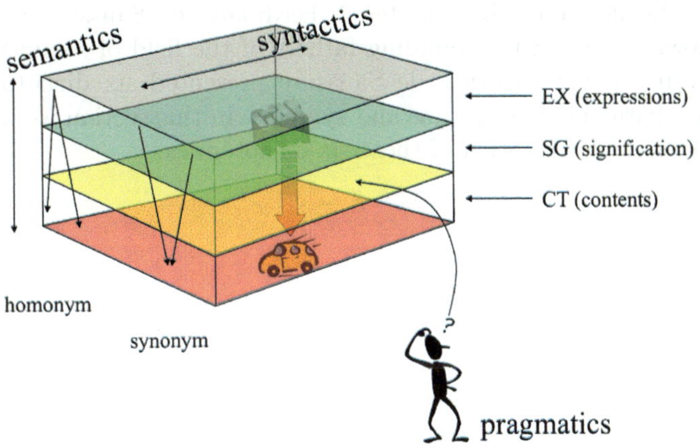

Fig. 2.5: The Semiotic Anatomy as suggested by Roland Barthes (1964). Note the syntax as the horizontal axis and semantics as the vertical axis. Note too: a synonym is a word that starts with the same content but is expressed differently (that is with a different expression) where as a homonym starts with the same expression but ends with a different content.

2.2.4 Barthes

Roland Barthes was one of the first scholars to expand the use of semiotics beyond the linguistic realm. As a cultural analyst, Barthes used semiotics to explain the meanings of behaviors and cultural symbols. Barthes didn't stop there, though. He extended the use of semiotics to gastronomy, fashion, sociology, and even traffic signs. He was truly the first individual to act on the idea that semiotics can be used to study everything. Barthes was largely responsible for the structuralist movement of semiotics (a movement based on Saussure's ideas). In his book *Elements of Semiology* [5], Barthes suggested an anatomy of semiotics (see Figure 2.5). This anatomy of semiotics allows us to show how semiotic elements interact across different axes. Syntactics, for example, are located on the horizontal axis and are defined as the juxtaposition of different signs (as in a sentence or musical phrase). Semantics, the process of choosing how to get from expression to meaning, is represented on the vertical axis. Finally, pragmatics, or the usage of signs, is represented by the signification component of the semiotic architecture. In this way, Barthes was able to make an anatomical model for semiology that allows it to be flexible enough to expand across different fields.

2.3 De Saussure's Six Dichotomies

Summary. As mentioned in Chapter 2, Ferdinand de Saussure was a Swiss linguist known as one of the founding fathers of the field of semiotics. In one of his most important theories, de Saussure presented six dichotomies that describe the characteristics of signs and symbols. In this section we will discuss and provide examples for each of these six dichotomies.

– Σ –

Here are the six dichotomies:

2.3.1 Signifier/Signified

The signifier/signified dichotomy deals with the differentiation between the expression and the content being represented. The signifier is an expression (e.g. musical notes in a score) that represents a certain meaning (e.g. pitch or rhythm). This is captured by the process of signification described in previous chapters, in which a sign goes through the stages of Expression, Signification, and Content. Music is often expressed by the surface of a score. This surface is full of expressions in the form of dynamics, written directions, and notation. Each of these are expressions (signifiers) that represent a musical meaning (the signified) such as rhythm, style, or pitch.

2.3.2 Arbitrary/Motivated

The arbitrary/motivated dichotomy has to do with the process of creating meaning. More accurately, it has to do with the way in which the expression is related to the content. If the expression is not related to the content, then it is said to be *arbitrary*. A sign is *motivated* if its expression is connected (that is analagous) to the content. This can be demonstrated by the case of *long play* (LP) records, also known as *vinyl records*, vs. compact discs (better known as CDs), see Figure 2.6.

The grooves in an LP disc are analogous to the musical sound (variable air pressure) that is produced. This is due to the fact that an electro-magnet in the head of the needle moves up and down in correspondence with peaks and valleys of the groove. This is translated into fluctuations in electrical current. These fluctuations cause a cone in the loudspeaker to vibrate, creating sound. This process is motivated, because the grooves in the LP disc directly influence the way in which the sound is created.

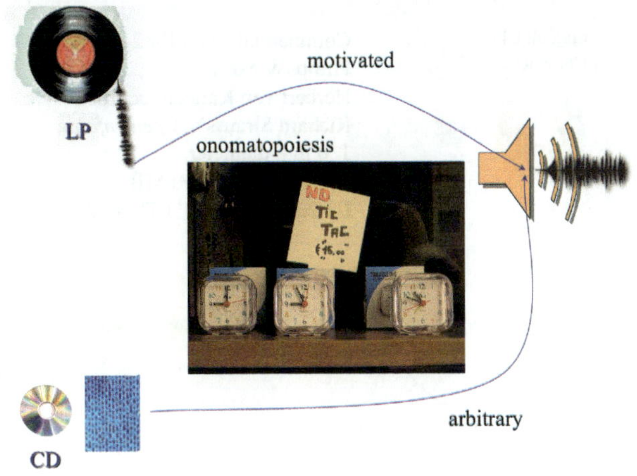

Fig. 2.6: An LP signifies its sound content in a motivated way, as opposed to a CD, whose digital expressions are arbitrary.

The surface of the CD is covered in bumps, which reflect light from a laser in the disc drive. This reflection contains binary information that is perceived by the laser lens, and decoded by the disc-driver or computer. The information is written in *binary*, a code consisting of 1's and 0's, in which 1 means "present" and 0 means "not present". Unlike the micro-grooves of the LP disc, which move up and down in correspondence with the encoded music, the binary information on the CD does not necessarily mirror the movement of the sound.

2.3.2.1 The Digital Approach, Sampling

Philips and Sony have used digital encoding since 1982 upon recommendation by Herbert von Karajan. It is remarkable that his acceptance of the CD quality (characterized by a 20 kHz upper frequency limit in the Fourier spectrum) was decisive for these companies, although Karajan's age (around 50 then) was not ideal as a reference for faithful sound perception. The hardware display of a CD is shown in Figure 2.7. The CD's laser reads the land and pit levels engraved upon the polycarbonate carrier as bit sequence of zeros and ones.

The transformation from the analog soundwave to the digital representation on the CD is explained in Figure 2.8. The wave is quantized in two ways: First, the sound's amplitude is quantized by 16 Bits[1]. This means that we are given $2^{16} = 65,536$ values $(0, 1, 2, \ldots 65,535)$ defined by the binary integer representation $b_{15}2^{15}+b_{14}2^{14}+\ldots b_1 2^1+b_0 2^0, b_i = 0, 1$. Second, the (quantized) values of the wave are only taken every $1/44,100$th of a second, i.e. the sample rate is $44,100$ samples per second. In total, this gives 635 MByte(1 Byte = 8 Bit)

[1] A one-Bit quantization would allow for just two values, 0 or 1.

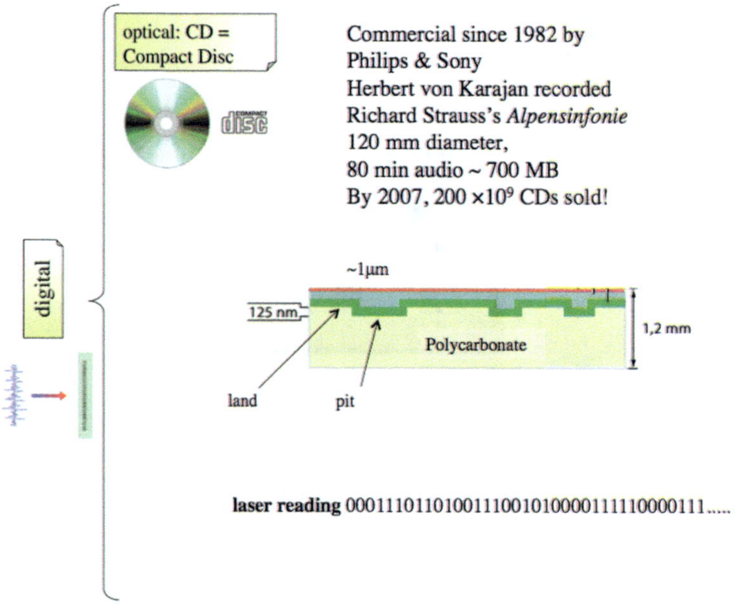

Fig. 2.7: The hardware display of a CD. The CD's laser reads the land and pit levels engrained upon the polycarbonate carrier as a bit sequence of zeros and ones.

CD capacity for one hour stereo recording. This allows for partials up to frequency $44,100/2 = 22.05$ kHz. The human ear is known to perceive up to 20 kHz.

Karajan, in his fifties, could very probably not hear more than 15 kHz, so for him the quality of the CD was perfect. A young human however would have asked for higher resolution. In recent technology, a sample rate of 96 kHz with amplitude quantization of 24 Bit is available.

This looks like the endpoint of a long development of sound conservation and transfer, for which the 200 billion CDs sold by 2007 is a good argument. However, the creative argument comes from the critical concept: the container unit of music, the CD. What is a wall thereof? First, its material part, this disc, why should this be the container? And second, why should music be transferred using such a hardware container?

Because the internet has defined the global reality of information transfer since the early 1990s, the answer to the above questions has become straightforward: Of course music data transfer can be accomplished via Internet, however, the speed of transfer is heavily dependent upon the FFT (Fast Fourier Transform) algorithm.

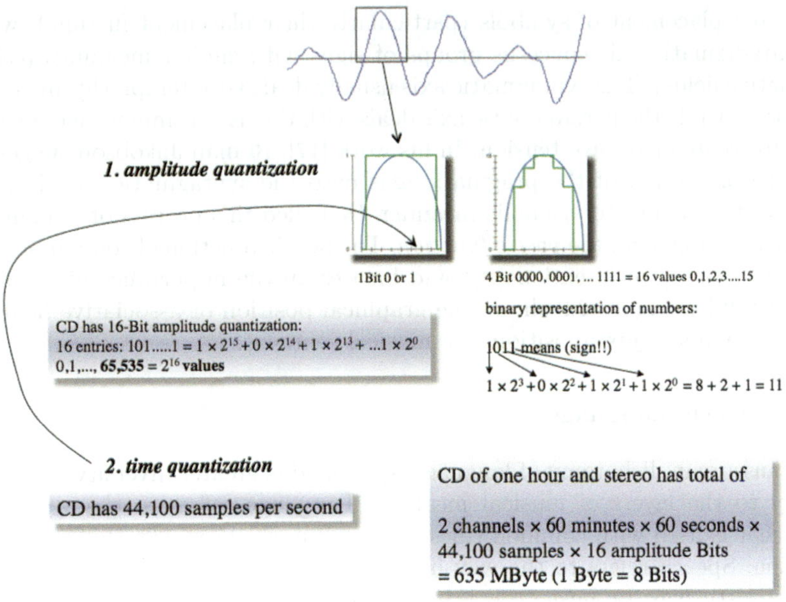

Fig. 2.8: The transformation from the analog soundwave to the digital representation on the CD.

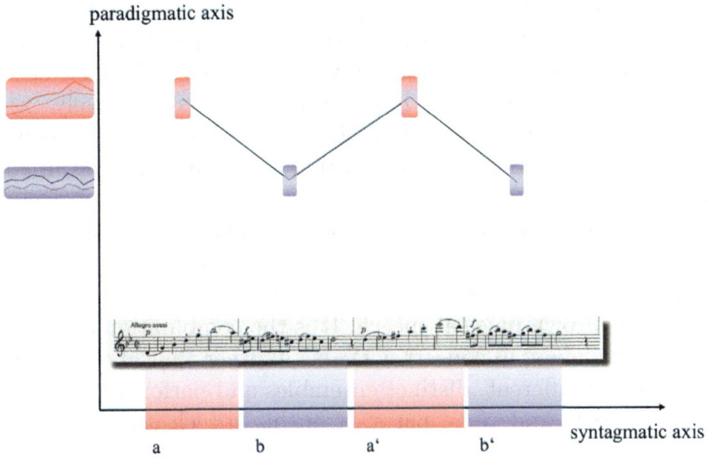

Fig. 2.9: Associative fields in their syntagmatic juxtaposition in a musical sentence.

2.3.3 Syntagm/Paradigm

Saussure's syntagm/paradigm dichotomy takes a central role in providing context to the semiotics of music. This dichotomy is based on two axes of meaning: the syntagmatic and the paradigmatic. The syntagmatic axis describes the position and placement of symbols (particularly their placement in time), while the paradigmatic axis concerns groups of signs with similar meanings (called associative fields). The syntagmatic axis usually deals with tempi, rhythms, and note order while the paradigmatic axis deals with the use of similar melodic patterns to create or resolve tension. In his work [17], Roman Jakobson suggested that the projection of the pragmatic axis onto the syntagmatic axis is what gives poetical depth to symbolic meaning. He called this process of (communication, art, language) the *poetic function.* The poetic function of communicative symbols is based in intentionality, and focuses on the importance of choosing signs carefully. Figure 2.9 shows the graphical position of associative fields in their syntagmatic juxtaposition in a musical sentence.

2.3.4 Speech/Language

The Saussurean dichotomy of language/speech adds greater diversity and complexity to the layers of musical meaning. Language refers to the system of signs that is used with common rules across all users. In music, this is musical notation. Speech refers to the way in which an individual uses language in a characteristic way. In music, this is representative of performance, or the music itself.

The conductor is given this creative license—to manipulate the group to ensure that a common language is understood throughout the whole ensemble. The language of the group, however, may be semi-dependent on the conductor's interpretation of the music. That is, the conductor's speech, his or her individual patterns of musicality, will influence what the ensemble must accept as a common language. The conductor, in this way, is able to creatively express his or her musicality through the ensemble. Despite this, each player still adheres to their own level of speech in that no other musician can play in the exact same way.

Soloists in particular are given an opportunity to express their musical speech, even to the extent of disregarding the conductor. Despite being most prominently observed in the conductor and soloists, all performers in the ensemble possess their own musical speech. It is the combination of these unique musical identities playing in concert with one another that gives rise to the incredible diversity of different, distinct ensembles and performances, even given the same instruments and vocal parts. It is this speech aspect of music that makes it at all possible to produce creative expression of individuality through music. Understanding the language/speech dichotomy in music is crucial for the creation of a meaningful interpretation.

2.3.4.1 Speech and Language Examples: Bach and Schönberg

In a semiotic system there are over-arching rules which define the social usage of the system. For example, grammar in a language system. There is also an individual usage of a semiotic system which may deviate from the over-arching

Fig. 2.10: Johann Sebastian Bach (1685-1750).

Fig. 2.11: Arnold Schönberg (1874-1951).

rules. For example, in the language system it is the individual composition of sentences. Or in gastronomy you have a recipe which is part of the system, but the re-creation of the dish itself is an individual expression.

In music, the language perspective is typically shared by reference to eternal values such as Pythagorean principles or the compositions by Johann Sebastian Bach (Figure 2.10). As a devout Christian, Bach's compositions reflect a desire to represent the beauty of an eternal system created by perfect deity. His music is known for immaculately following the rules of music theory such as counterpoint. This adherence to pre-defiend rules makes Bach and his compositions a perfect example of the concept of system.

In opposition to this eternal system, Arnold Schönberg (Figure 2.11) invented an individual human approach to composition, the twelve-tone method, in 1923. Schönberg's music is largely atonal, and often not especially pleasant, especially for those who are used to listening to consonant chords. Schönberg's divergence from the normative standards is an excellent example of Saussure's concept of speech. Schönberg, unsatisfied with the pre-existing rules of music theory, created his own language in which to express himself as a creator of new rules.

Music in the twelve-tone scale has received mixed feedback. Many people simply find the music unpleasant to listen to, an attitude that was felt strongly by many critics as well. Yet, others supported Schönberg, considering his new system to be a work of genius. You can decide for yourself whether or not Schönberg's music is good, but for the purposes of this book it is important to note that his creation of an entirely new composition style is emblematic of the concept of speech.

2.3.5 Synchrony/Diachrony

Saussure suggested that a semiotic system involves symbols that are distributed in both time and space (Figure 2.12). To account for this, Saussure proposed

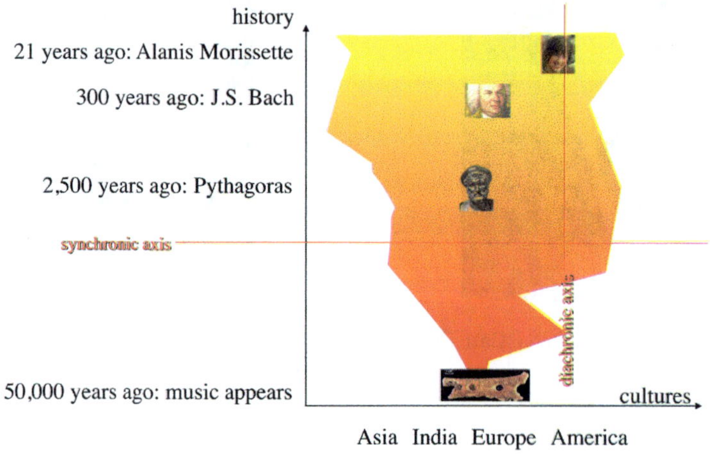

Fig. 2.12: The diachronic and synchronic axes of music semiotics.

the synchrony/diachrony dichotomy, which acts as a sort of coordinate system. Here we will use it to identify music by its time and place in history.

The synchronic axis focuses on "place" and represents what is happening music across all forms at any one time. The units of this axis are categorical, focusing on classifications of musical form such as culture or genre. It deals with music ethnology or ethnography. The diachronic axis describes the development of a form of music over time. The units of the diachronic axis are measured in time (usually in years). Typical structures of diachronicity are the history of words, i.e., their etymology.

2.3.6 Lexem/Shifter

For every word there is an objective definition. However, the meaning of a word may change due to context. For example, if I say "you" it has a different meaning than when you say "you". When I say "you", I am referring to you, whereas when you say "you", you are referring to me. This rather confusing conundrum is the concept from which the lexem/shifter dichotomy is born.

Lexem refers to signs that can be objectively defined, while shifters are signs that change meaning depending on the context of their users. In music, the majority of symbols can be considered shifters. This makes interactions with music highly dependent on context and interpretation.

Since the composer cannot write every intention into the score, he or she relies on the performer to bring out the meaning that cannot be derived from the symbols on the page. It is the task of the musician to add a human element, an element of creativity, as an interpretation of the score. This individuality is expressed through speech (as discussed previously). See Section 2.3.6.1 for an example.

2.3.6.1 Semiotics in Music Performance: the Example of Celibidache's Ideas

Fig. 2.13: Sergiu Celibidache (1912-1996).

For this section of the book, we would like to recommend that you go to YouTube and search for Sergiu Celibidache (Figure 2.13) conducting Gabriel Fauré's *Requiem* with the London Symphony Orchestra. The videos not only exemplify the concepts of semiology that we have heretofore discussed, they also provide unique insights and explanations from one of the most famous conductors in European history.

Sergiu Celibidache can certainly be called a man who was stuck in his ways. Celibidache considered music as a form of spirituality, and was quite stringent in his reverence for it. In fact, he generally would not allow his music to be recorded, as he believed this removed the authenticity of it. In his own words, Celibidache considered recordings "a standardization, a dehumanization of the reaction [music]". He was extremely vocal about the importance of music as an expression of individuality, as well as an excellent director who was able to communicate to all of the members of an ensemble in order to coordinate the magnificent power of a symphony.

As you may have noticed already in this short explanation alone, Celibidache's assertions are highly reminiscent of the Saussurean dichotomies. In particular, Celibidache seems to emphasize the importance of the language/speech and lexem/shifter dichotomies. He focuses on how music is simultaneously an expression of individuality (speech) and a comprehensive group effort to interpret the piece (language). In reference to the lexem/shifter dichotomy, Celibidache emphasizes the importance of the "shifter" quality in music performance. In fact, Celibidache suggests that music performance has a shifting nature.

Let us begin with the language/speech dichotomy. As you remember, language refers to the system of symbols that is used with common rules across all users. In music, this is staff notation. Speech refers to the way in which an individual uses language in a characteristic way. In music, this is representative of individual interpretations, performances, and the way in which the music is realized. In the videos, Celibidache often stops his ensemble to correct articulation or pronunciation. For example, in Faure's *Requiem*, he stopped his ensemble for their pronunciation and articulation of the phrase "Sanctus Dominus", saying "Not on the spot, please. Sanctus DoOMinus, in-between, not on DOMinus." In this way Celibidache was working on establishing common language rules for his choir, ensuring that the collective group was performing in the same, desirable way.

In an ensemble, the conductor is has such a creative license—to guide the group's performance to ensure that a common language is understood throughout the whole ensemble. The language of the group, however, is usually dependent on the conductor's interpretation of the music. Thus, the conductor's speech, his or her individual patterns of musicality, will influence what the ensemble must accept as a common language. The conductor, in this way, is able to creatively express his or her musicality through the ensemble. Despite this, each musician still adheres to their own level of speech in that no other musician can play in the exact same way.

However, there are some special cases in which the speech of an individual is allowed to override the language of the entire group. Soloists, for example, are given the freedom to express their musical speech, even to the extent of leading the conductor. In explanation of a soloist's opportunity for self-expression, Celibidache said "Away from your possibilities, not out—not for me—but for yourself is a little bit better" thus explaining to the soloist that he was free to express some artistic interpretations.

To the soprano soloist in the Fauré's *Requiem* Celibidache said "You are the queen, we have to bow. The question is not to be on time, because the time is not there. You create the time." In this example, Celibidache illustrates the power of musical speech, the freedom to express with your own character whatever artistic interpretation of the music that you hold.

Despite being most prominently observed in the conductor and soloists, all performers in the ensemble possess their own musical speech. It is the combination of these unique musical identities playing in concert with one another that gives rise to the incredible diversity of different, distinct ensembles and performances, even given the same instruments and vocal parts. Through speech, it is possible to produce individual creative expression. Understanding the language/speech dichotomy in music is crucial for the creation of a meaningful interpretation.

In regards to the lexem/shifter, dichotomy of music, Celibidache emphasizes the importance of live music. In the semiotics of music, the majority of signs can be considered shifters by nature. This makes interactions with music highly dependent on context and individuality. In reference to the use of a score, Celibidache says "This is why he wrote the text—which is unreliable and uncomplete—we must find what is not written in the score." To Celibidache, the importance of the shifter state of music is quite clear. Since the composer cannot write every intention into the score, he or she relies on the performer to bring out the meaning that cannot be derived from the symbols on the page.

In addition to the interpretative dimension added by the performer, Celibidache asserts that the meaning of music is affected by the performance itself, including the tempo, acoustics, and location. For this reason he never openly condoned the act of recording musical performances. "The reason I never made a recording is that I never found anyone who could make a recording. Those are photos! Photos of a reality that cannot be photographed. Do you play the record in the acoustics where they have been taken? The acoustics of a concert hall has such an importance—the tempo depends on it!" Celibidache continues to argue that the process of recording is a dehumanization of the music, taking away much of the character that the musicians work to create. To Celibidache, musical creativity not only depends on the importance of understanding the separation of music from musical notation, but also the understanding that music is shaped largely by its context. Without a contextual setting, we cannot experience music in the same way, it loses a part of its meaning.

2.4 The Babushka Principle in Semiotics: Connotation, Motivation, and Metatheory

Summary. In this chapter, we introduce the idea of what we call the *semiotic Babushka Principle* (sign systems within sign systems) presented by Louis Hjelmslev. We discuss the implications of applying this principle to the analysis of music and musical scores. We argue that, through the conceivably infinite mapping of connotational systems, music is capable of accomplishing significant symbolic depth.

– Σ –

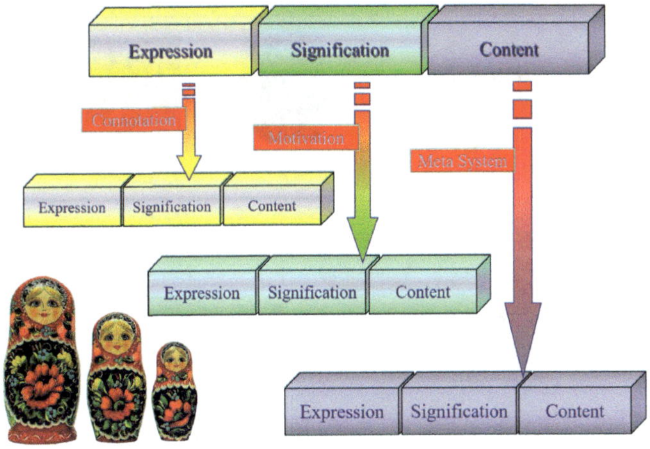

Fig. 2.14:

jelmslev's ramifications of a semiotic system into connotation, motivation, and metasystem.Babushkaht As we have seen in Chapter 2.1, expression, signification, and content are the basic components of meaningful signs and symbols. However, it is somewhat ambiguous which part of a signs corresponds to which step in the process. For example (Figure 2.15), you might say that the expression of the word "firework" is the written word itself, the signification being the understanding of the word, and the content being noise, light, and color. However, your friend may suggest that the noise and color associated with firework are the expression, where as a connotative meaning such as "Fourth of July" may make up the content.

In reality, neither of you are wrong. Rather, you are operating on different levels of symbolism. The role of connotation, brought up by your hypothetical friend (we realize that your friends probably don't casually have conversations about semiotics) was actually first brought up by Louis Hjelmslev. Hjelmslev's Babushka Principle is the idea that, through connotation, one part of a sign system can become a sign system of its own, resulting in a possibly infinite system of signs systems within sign systems.

The Babushka Principle is exemplified by the expansion of the original three components to a sign system—expression, signification, and content (see Figure 10.1 for a visual representation). An expansion of the expressive dimension is called *connotation*, while expansion in the signification dimension is called *motivation* and in the dimension of content is called *meta system*. Figure ?? shows these ramifications, and Figure 2.15 shows the connotational double articulation in language.

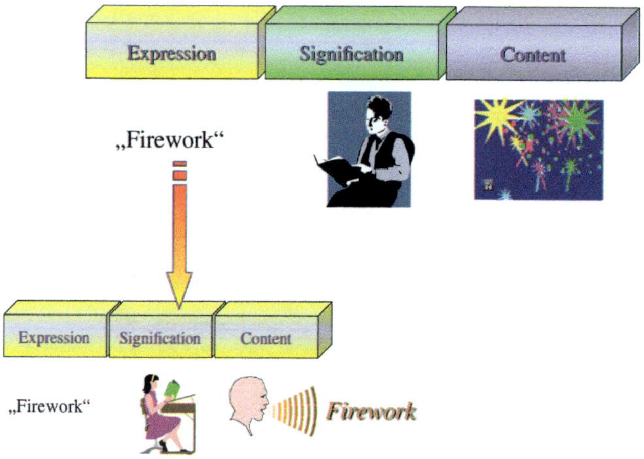

Fig. 2.15: Double articulation in language, a connotative structure. It extends Figure 2.1.

For example, let us look at how a musician perceives, analyzes, and performs a score, see Figure 2.16. The first sign system that we may conceive is that consisting of the score's text, the reading of the score, and the conception of the score by the composer (expression, signification, and content respectively). However, this entire sign system only accounts for the perception of music, ignoring the importance of analysis and performance. The semiotic nature of music does not end there, but the first sign system in itself is complete.

In order to solve this issue, we must also be aware of a second sign system involved with score analysis, including the conception of the score, the musician's analysis, and the form the music takes in the composer's mind after analyzing it (again, expression, signification, and content respectively). This sign system is a connotational system of the original sign system. That means that it takes the content of the first system (i.e. the conceived score) as an expressive level, and expands it into a sign system of its own. Similarly, we can say that the original sign system is a connotation of the second sign system, because it takes the expression (i.e. the conceived score) and expands it into a sign system of its own.

This process of condensing and expanding sign systems is responsible for the translation of the score into a performance, and ultimately to the listener's interpretation of the performance (see Figure 2.16). One can even argue that sign systems underly the way in which a society views a piece of music or a certain performer. Such an assertion is reminiscent of the psychological concept of schema formation, in which individuals form an understanding of something new through activating an intricate, underlying web of connections of related concepts. The convergence of these theories lends support to their validity.

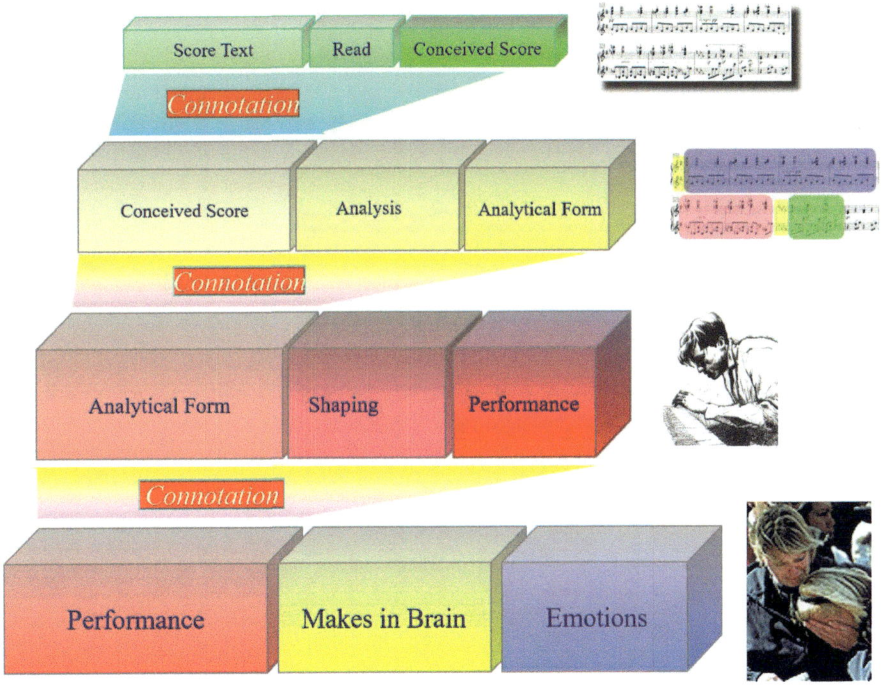

Fig. 2.16: The multiply connotational imbrication of a musical performance.

2.4.1 The Structural Consequences of the Babushka Principle

The Babushka principle extends the classical horizontal direction from expression through signification to content in a vertical direction of connotation, motivation, and metatheory. Semiotics is in this abstract sense a two-dimensional display.

This means that a semiotic localization can be extended horizontally and vertically by first determining its horizontal coordinates

(Expression/Signification/Content),

and then its vertical coordinates

(Connote/Motivate/Metatheorize).

This is the structural base of the following mathematical approach to a precise description of semiotical anatomy.

3

Functorial Semantics Category

Summary. This chapter presents the category of functorial semantics, which formalizes the semiotic objects together with the connecting morphisms. These objects are called H-jets, "H" standing for Hjelmslev, who introduced the vertical dimension in a comprehensive semiotic.

$- \Sigma -$

3.1 The Overall Construction

The general plan is this. We first introduce the basic objects of semiotics: signs. Then we define a big digraph $\mathcal{LD}$ ($\mathcal{L}$ for "Limit", $\mathcal{D}$ for "Digraph") that is generated by signs and their Hjelmslev perspectives (connotation, motivation, metatheory), together with the sign relationships. We then define a category $\mathcal{LDC}$ ($\mathcal{C}$ for "Category") that is derived from these digraph structures. Finally, we define more subtle structures of motivation that specify our category $\mathcal{LDC}$, which is core to the semiotic system we want to investigate.

3.1.1 The Basic Digraph

The semiotic we are envisaging gets off the ground with its basic objects: signs. A sign X is a triple $Ex(X), Sg(X), Ct(X)$ that consists of these objects: the sign's *expression* $Ex(X)$, its *signification* $Sg(X)$, and its *content* $Ct(X)$. So far, these components can be any entities, objects, actions. We represent this triple component display graphically by

$$\begin{array}{ccccc} Ex(X) & \dashrightarrow & Sg(X) & \rightsquigarrow & Ct(X) \\ & \diagdown & | & \diagup & \\ & & X & & \end{array}$$

We shall now define a digraph $\mathcal{LD}$ (the limit digraph) by the following components.

G. Mazzola et al., *Functorial Semiotics for Creativity in Music and Mathematics*, Computational Music Science, https://doi.org/10.1007/978-3-030-85190-3_3

- The digraph's *vertices* are what we call *H-jets*, H standing for Hjelmslev. These are sequences $x_. = (\dots x_{i+1}, x_i, \dots)$ of integer-indexed objects that may be finite or infinite to the left and to the right. The instances are connected as follows. A couple x_{i+1}, x_i can be either $x_{i+1}, x_i = Ex(x), x$ or $x_{i+1}, x_i = Sg(x), x$, or $x_{i+1}, x_i = Ct(x), x$ for a sign x. There is therefore no H-jet with only one instance. Such sequences are all connecting signs to their components. We usually represent such H-jets graphically as vertically oriented lines:

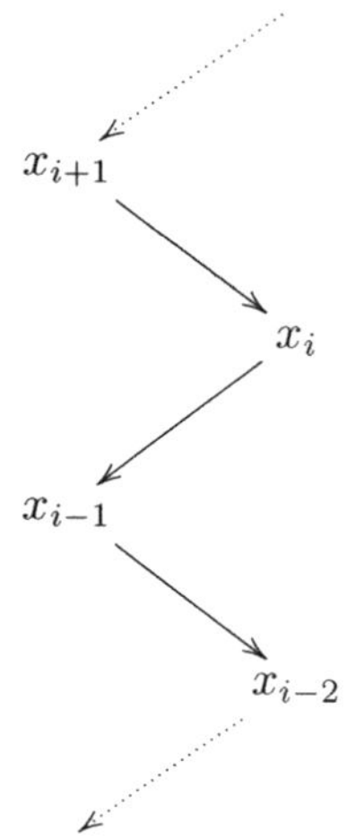

For example, the double articulation in language states that the acoustical content $Ct(X)$ of a graphical word sign X is the expression $Ct(X) = Ex(Y)$ of the logical content $Ct(Y)$. Or else, the biological content Ct(Albatros) of the word "Albatros", the type of bird, in Baudelaire's poem *L'Albatros*, is the expression of the poetical concept of freedom. We may then combine these two constructions to yield a H-jet that reaches from the graphical word expression "Albatros" to the poetical concept content of freedom.
It is essential to view these vertices as semiotic entities by their own, each H-jet represents a determined sign structure. In particular, if a H-jet $x_.$ is a subsequence of another H-jet $y_.$, this does not mean that it can be absorbed by the supersequence. It is, semiotically speaking, an autonomous object. This extension of ordinary sign components is crucial when we want to capture the totality of a sign's ramifications. Humans effectively operate on this level in their semiotic processing. We claim that the original structure of a sign (aliquid pro aliquo) is only a reduced and far too abstract aspect of semiosis. Abstraction meaning that one *pulls away* what matters: the Latin origin *ab-trahere* literally hits this reductionist activity. Our wording "limit" in the digraph $\mathcal{LD}$ stems from the idea that the digraphs H-jets can evidently be seen as projections of limit structures, as 'pre-liminary' entities.

Definition 1. *We call a H-jet* $Y = (y_0, y_1)$ *with the minimum of two coordinates* elementary.

It is also important to understand that a H-jet $x_.$ may have proper self-intersections, i.e., it can happen that $x_i = x_j$ for $i \neq j$, graphically looking like this:

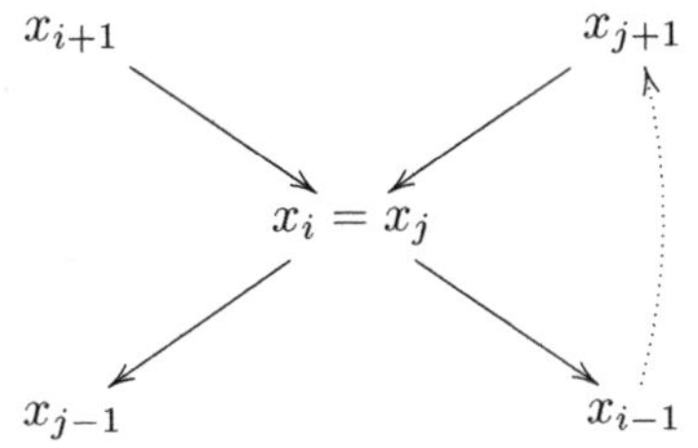

- Given two vertices $x_.$ and $y_.$ of $\mathcal{LD}$, an *arrow* $x_. \to y_.$ is defined in two situations:
  1. we specify two indices i_1 of $x_.$ and i_2 of $y_.$, we then suppose that $x_{i_1} = y_{i_2} = z$; we further suppose that there are two elements x_{i_1+1}, y_{i_2+1} and that either $x_{i_1+1} = Ex(z)$ and $y_{i_2+1} = Sg(z)$:

     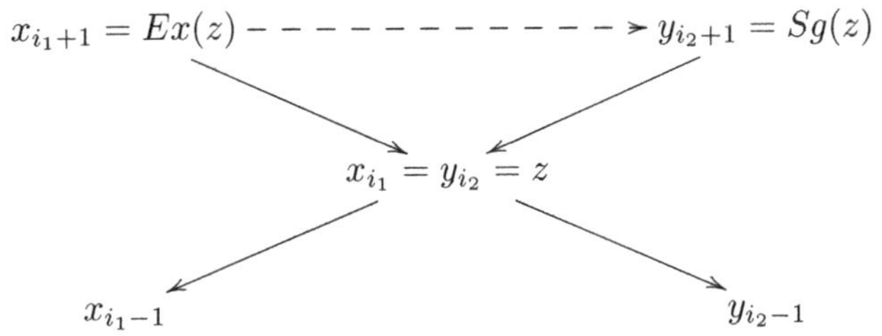

     or $x_{i_1+1} = Sg(z)$ and $y_{i_2+1} = Ct(z)$:

     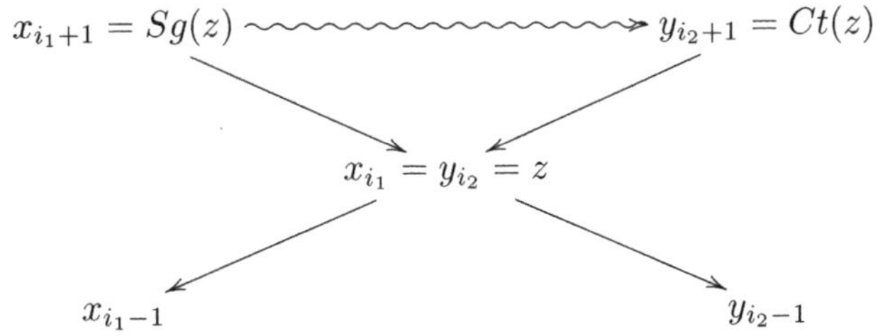

     Observe that the specification of the two indices i_1, i_2 is part of the definition of such a "crossing" arrow.
  2. the vertex $x_.$ is essentially a sub-H-jet of $y_.$. More precisely: there is a map $f : i \mapsto f(i)$ on the index set of $x_.$ into the index set of $y_.$ that is injective and has no gaps, i.e. $f(i+1) = f(i) + 1$ and $x_i = y_{f(i)}$ for all indices i of $x_.$.

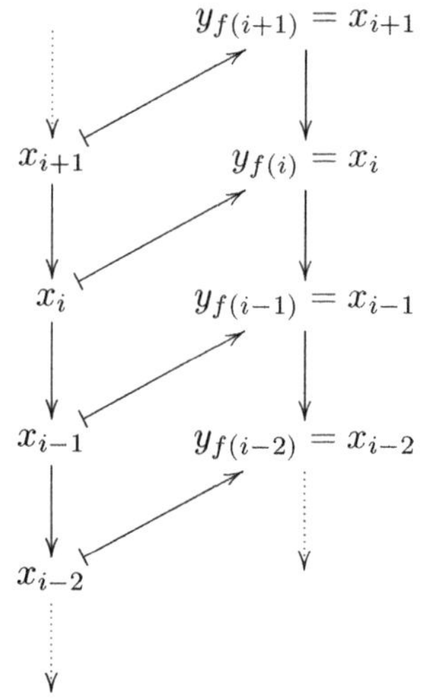

Observe that there might also exist loops, i.e., arrows $x_. \to x_.$ in $\mathcal{LD}$. For the first arrow type, this looks as follows, for example:

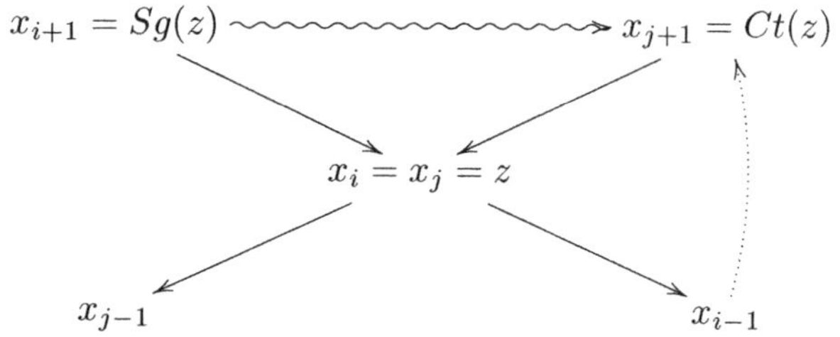

For the second arrow type, it means that $x_.$ has a periodic value structure.

Lemma 1. *If $x., y.$ are two H-jets containing a sign $z = x_i = y_j$ at their top indices i and j, then there are extensions $x^*_., y^*_.$ of $x., y.$, respectively, such there is a morphism $x^*_. \to y^*_.$*

In fact, we may extend $x_.$ at the top by $Ex(x_i)$, and $y_.$ at the top by $Sg(y_j)$ and then take the morphism that is defined by this new top configuration.

3.1.2 The Basic Category

Starting from the (big) digraph $\mathcal{LD}$, we consider its path category $Path(\mathcal{LD})$. The arrows stemming from sub-H-jet inclusions however should admit their composition without generating new paths. Therefore, we define the category $\mathcal{LDC}$ to be the quotient $\mathcal{LDC} = Path(\mathcal{LD})/R$, where R are the relations defined by composition of sub-H-jet injections as well as composition of type $f : X \to Y' \subset Y$, where $Y' \subset Y$ with identical indices, and $X' \subset X \to Y$, where $X' \subset X$ is a sub-H-jet with identical indices such that the morphism $X \to Y$ connects elementary crossing parts (of an arrow of type (1)) within X'. In particular, a renumbering of a H-jet defines an isomorphism in $\mathcal{LDC}$.

Often, to keep notation simple, we shall use the word "sign" to refer to a H-jet. It should always be clear from the context, what we mean.

3.2 Quotient Categories

The big category $\mathcal{LDC}$ is essentially the path category of digraph $\mathcal{LD}$. Semiotics does however suggest and/or admit a number of more concrete categories that are defined as quotient categories $\mathcal{LDC}/R$ for relations R on the morphisms of $\mathcal{LDC}$. For example, suppose that a sign X is "equivalent" to a sign Z, i.e., we have a pair of sign structures

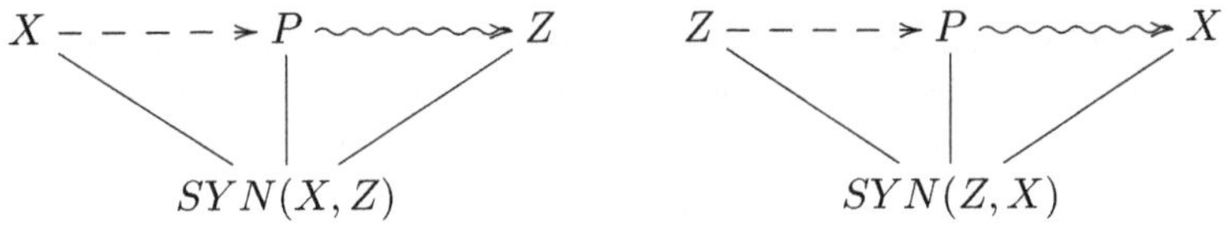

(The signification P is the same to keep things simple. One could also envisage two synonymies: one for identical contents, and one for identical significations, following Frege's distinction of Bedeutung (significate) versus Sinn (sense)) Then the composition (in both directions, using the arrow given by the sign components of X and Z, respectively) of the two morphisms defined by these signs could be asked to be the identity on X and on Z, respectively. This would define equivlence as an isomorphism of signs. A second example of a relation that one could envisage is given by two arrows: one, $i : Y \rightarrowtail Z$, which is the embedding of sub-H-jet Y in Z. The other, $f : Z \to Y$, is given by this configration (the dotted arrow from z_1 to y_3 shows the inclusion of Y in Z):

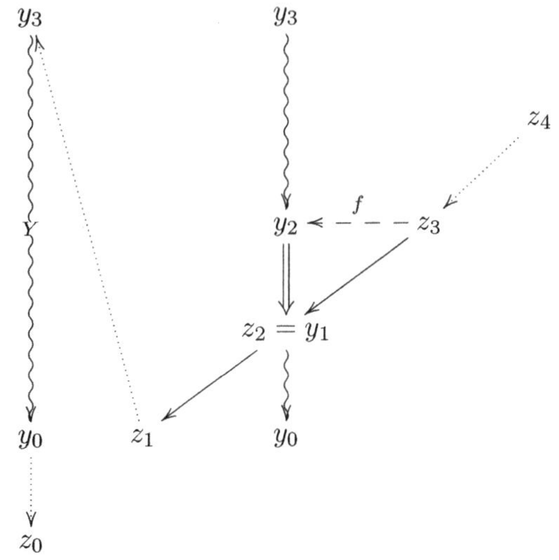

One can ask for the composition $f \circ i$ to be the identity Id_Y, but this is not mandatory from a semiotic point of view.

One could also step over to the category of fractions $\mathcal{LDC}/R[\Sigma^{-1}]$ defined by a set Σ of morphisms of $\mathcal{LDC}/R$ that are to become isomorphisms [11]. We shall denote such a quotient category by $\mathcal{LDC}^{\Sigma}_{R}$ to ease notation. To be clear, this needn't be admitted since such an equivalence can also generate different contents that are associated with different naming conventions.

Let us order relations R, S and sets of morphisms $\Sigma \subset \mathcal{LDC}/R, \Xi \subset \mathcal{LDC}/S$ according to their inclusion $R \subset S$, with the corresponding functors $/R/S : \mathcal{LDC}/R \to \mathcal{LDC}/S$, and the condition $/R/S(\Sigma) \subset \Xi$. Then we get a hierarchy of quotient categories of the basic semiotic category $\mathcal{LDC}$, defined by the canonical functors

$$\mathcal{LDC}^{\Sigma}_{R} \to \mathcal{LDC}^{\Xi}_{S}.$$

It is obvious that this theory should also consider proper subcategories of $\mathcal{LDC}^{\Sigma}_{R}$, together with functors between such subcategories. In the following discourse, we shall not stress this option, but it is understood that many of its techniques apply to subcategories as well.

3.3 Semiotics for the Yoneda Lemma

The Yoneda Lemma states (among others) that the functor $Yoneda : \mathcal{LDC}^{\Sigma}_{R} \to (\mathcal{LDC}^{\Sigma}_{R})^{@}$ is fully faithful, more precisely, we have an isomorphism $X@F = F(X) \stackrel{\sim}{\to} (@X)@F$ for any object X of $\mathcal{LDC}^{\Sigma}_{R}$ and presheaf F in $(\mathcal{LDC}^{\Sigma}_{R})^{@}$. This may be interpreted as a statement about a *global content*: To classify a sign X (a H-jet, to be precise) it is sufficient to know its functor $@X$, i.e., how it looks like when viewed from all other signs (H-jets, to be precise). This functor $@X$ defines the *global expressivity* of X. Alternatively, we may consider the covariant functor $X@$ which may appear even more as a representative of *global content* since it evaluates to all morphisms with domain X, which is a generalization of the sign content relationship $Ex(X) \to Ct(X)$.

Definition 2. Given a relation R of $\mathcal{LDC}$ among morphisms and a set Σ of morphisms in $\mathcal{LDC}/R$, the global *contravariant or covariant content* of a sign X in $\mathcal{LDC}^{\Sigma}_{R}$ is the presheaf $@X$ or $X@$ over $\mathcal{LDC}^{\Sigma}_{R}$.

The Yoneda method opens up a semiotic perspective that challenges our conceptualization of signs. In fact, since the category $\mathcal{LDC}^{\Sigma}_{R}$ is not co-complete[1], we encounter presheaves F in $\mathcal{LDC}^{\Sigma @}_{R}$ that are not representable. These functors define a new type of signs: They have a global content, but do not stem from signs or H-jets of signs. Intuitively speaking, such functors generate a meaning without meaning *something*, some sign. In Yoneda's functorial setup, the verb "meaning" has mutated to a non-transitive verb.

[1] Recall that every presheaf is a colimit of representable presheaves.

A straightforward method to evaluate such functorial presheaves is to look at natural transformations $n : F \to @Y$ for H-jets Y and the evaluating the maps $X@n : X@F \to X@Y$ to get normal morphisms with "contents" Y. The transformation n here plays the role of a kind of "content arrow" of F in Y. This idea means that we are transporting the morphism setup of content production between H-jets to the level of presheaves.

This can in fact help understand the functorial situation. The non-representable presheaves F admit a very natural evaluation of their functorial content $F@(@Y)$ for H-jets Y. We can use the fact that we have $F \overset{\sim}{\to} \mathrm{colim}_\iota @X_\iota$ for a diagram (X_ι) of H-jets. Just take the canonical diagram $\int F$, the category of elements [36, p. 43] of F. Then we have $F \overset{\sim}{\to} \mathrm{colim}(\int F \overset{p}{\to} \mathcal{LDC}^{\Sigma}_{R} \overset{Yoneda}{\to} \mathcal{LDC}^{\Sigma @}_{R})$, where p is the canonical projection to the first factor X of an element (X, x), and *Yoneda* is the Yoneda embedding. Take this colimit and call it $\mathrm{colim}_\iota X_\iota$. This entails that $F@(@Y) \overset{\sim}{\to} \mathrm{colim}_\iota X_\iota(@Y) \overset{\sim}{\to} \lim_\iota (@X_\iota)@(@Y) \overset{\sim}{\to} \lim_\iota X_\iota @Y$. The equation

$$F@(@Y) \overset{\sim}{\to} \lim_\iota X_\iota @Y$$

means the following:

Proposition 1. *The functorial content of a presheaf F in H-jet Y is the limit of a diagram of 'objective' contents $X_\iota @Y$ for a diagram of H-jets (X_ι).*

This enables us to calculate the objectivized content $F@Y := F@(@Y)$ at object Y when evaluating a functorial meaning in $X@F$. This is defined by a map

$$q : X@F \times F@Y \to X@Y.$$

In fact, recall that colimits of preseheaves evaluate pointwise. Then q takes a representative v_{ι_0} of an element v of $X@F$ and a family (u_ι), element of $F@Y$, to the morphism $u_{\iota_0} \circ v_{\iota_0}$. It is immediate that this definition is independent of the selected representative of v.

3.4 Constructing Colimits of Representable Functors

In this section, we want to work out the structure of a diagram (X_ι) of signs in order to create colimits of diagrams, which are the general case of presheaves over $\mathcal{LDC}^{\Sigma}_{R}$.

To begin with, we want to define a sign structure for functions $f : A \to B$ on sets. This looks as follows:

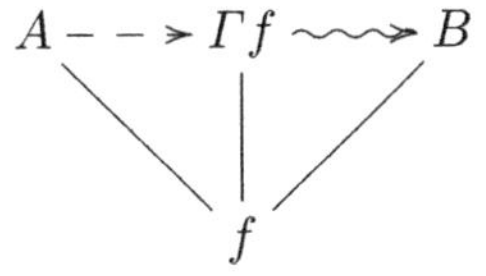

meaning that the domain of f is $Ex(f)$, the codomain of f is $Ct(f)$, and the graph Γf is $Sg(f)$. The latter is viewed as a mathematical gesture on the digraph ΓA that consist of copies $\uparrow_a$ of the arrow digraph, one for every element $a \in A$. The curve for $\uparrow_a$ is the arbitrary sign morphism $a \dashrightarrow P \rightsquigarrow f(a)$ that corresponds to the pair $(a, f(a))$ of Γf. Regarding relations R for quotient categories, it is evident that we would include the composition law of functions, i.e., the composition of two morphisms $A \dashrightarrow \Gamma f \rightsquigarrow B, B \dashrightarrow \Gamma g \rightsquigarrow C$ would be equivalent to $A \dashrightarrow \Gamma(g \circ f) \rightsquigarrow C$.

With this construction, a function $g : Fun(F_1) \to Fun(F_2)$ between two domains of identifiers IF_1, IF_2 of forms F_1, F_2, respectively, is realized as a morphism $v \circ v' \circ u' \circ u$ according to this diagram:

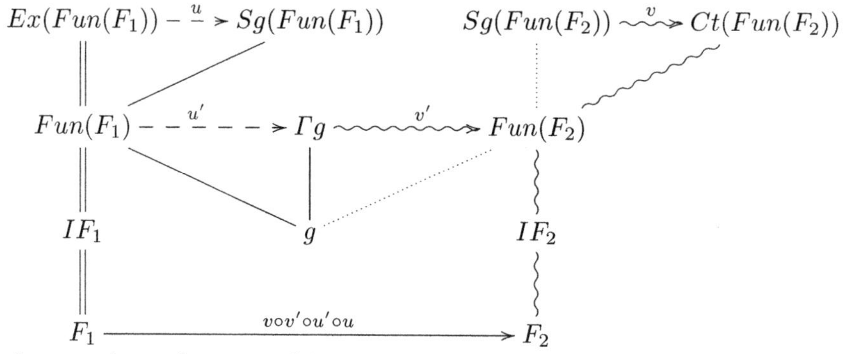

where we have these morphisms:

$$u : (F_1, IF_1, Fun(F_1), Ex(Fun(F_1))) \to (g, Fun(F_1), Sg(Fun(F_1))),$$

$$u' : (g, Fun(F_1), Sg(Fun(F_1))) \to (g, \Gamma g),$$

$$v' : (g, \Gamma g) \to (g, Fun(F_2), Sg(Fun(F_2))),$$

and

$$v : (g, Fun(F_2), Sg(Fun(F_2))) \to (F_2, IF_2, Fun(F_2), Ct(Fun(F_2))).$$

This type of semiotic morphism construction to represent morphisms between forms is used to design diagrams of forms. One should be aware of the possibility of defining directions of morphisms between H-jets *as desired* following the above method for morphisms u, v. Consider in fact four H-jets intersecting at sign z as follows:

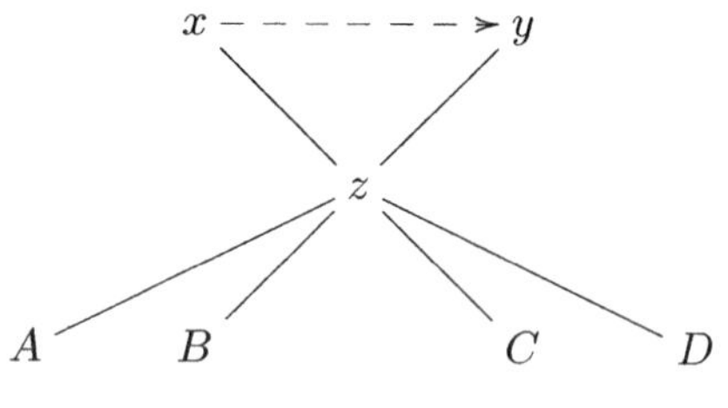

The horizontal top arrow (which is an Ex to Sg arrow here) can define a morphism in any direction for the H-jets that terminate in A, B, C, D. For example, we can have a morphism $(D, z, x) \to (B, z, y)$. The direction of the morphism is given by the combination of the H-jet parts below z and their extensions beyond z. Observe that the top horizontal arrow can also be a morphism that is composed by the $Ex \to Sg$ and the $Sg \to Ct$ parts of sign z, where the middle H-jet is just $(z, Sg(z))$.

This method enables us to define quite general diagrams of H-jets, and in particular diagrams of functors associated with forms for denotators. Consequently, we may now attempt to make explicit some functors F qua colimits of diagrams of H-jets. The basic idea here is to connect H-jet parts below z and above z in two ways, $=$ and $-$ connections, as shown in the following diagram, such that the resulting morphisms f, which result from the arrow $x \to r$, look in opposite directions.

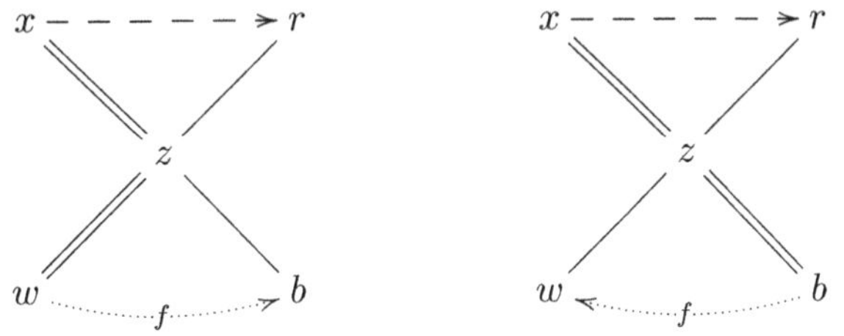

In a more complex context, defining morphims $A \to B \to C \to D \to E \to F$ in the category $\mathcal{LDC}$, this looks as follows, where the connections with same line type define parts of H-jets, and where the arrows $13 \to 14, 1 \to 2, 15 \to 16, 3 \to 4, 17 \to 18$ define the morphisms between these H-jets. Here the labels A, B, C, D, E, F are associated with parts of the H-jets A, B, C, D, E, F at the bottom of the diagram:

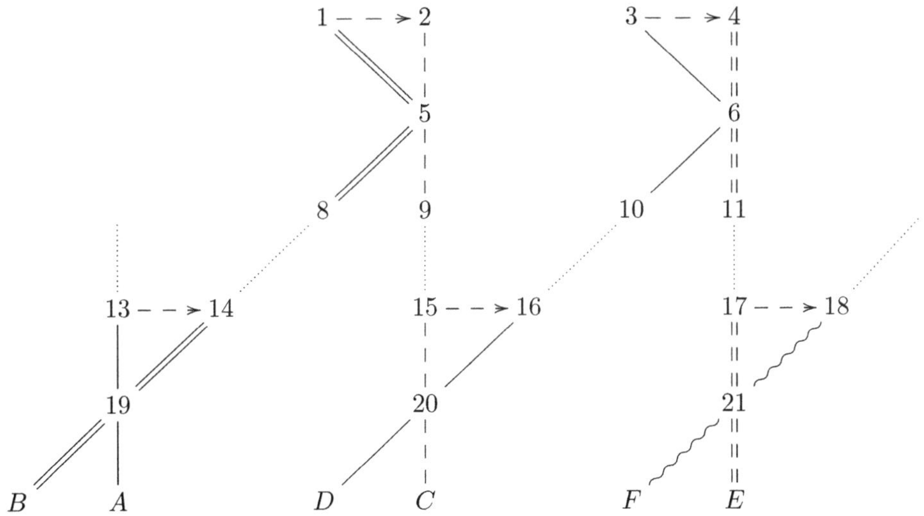

Let us now show how the colimit of the diagram $(@X_\iota)$ of H-jet presheaves as shown above looks like when taking the morphisms with domain H-jet X. In the next digram, we show as an example a number of morphisms $X \to B, X \to C, X \to D, X \to E$, which would have to be compatible with the colimit diagram if they were representing the same morphism $X \to \mathrm{colim}_\iota(@X_\iota)$. Recall from the Yoneda Lemma that $X@\mathrm{colim}_\iota(@X_\iota) = \mathrm{colim}_\iota X@X_\iota$.

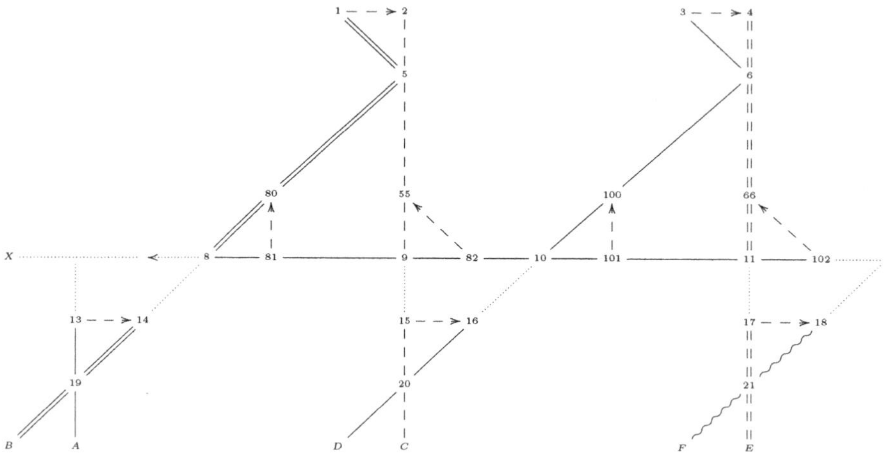

Reversing the system of morphism arrows $X \to B, \ldots X \to E$ which are given by the arrows $\ldots 81 \to 80, \ldots 102 \to 66$, one is led to consider a sequence of diagram-compatible morphisms $A \to Y, B \to Y, \ldots E \to Y$ which defines an element of $\mathrm{colim}_\iota @X_\iota \to Y$ (the universal property of the colimit). The following diagram shows one such composed morphism $f : X \to B \to Y$ that pertains to the composition $X \to \mathrm{colim}_\iota @X_\iota \to Y$.

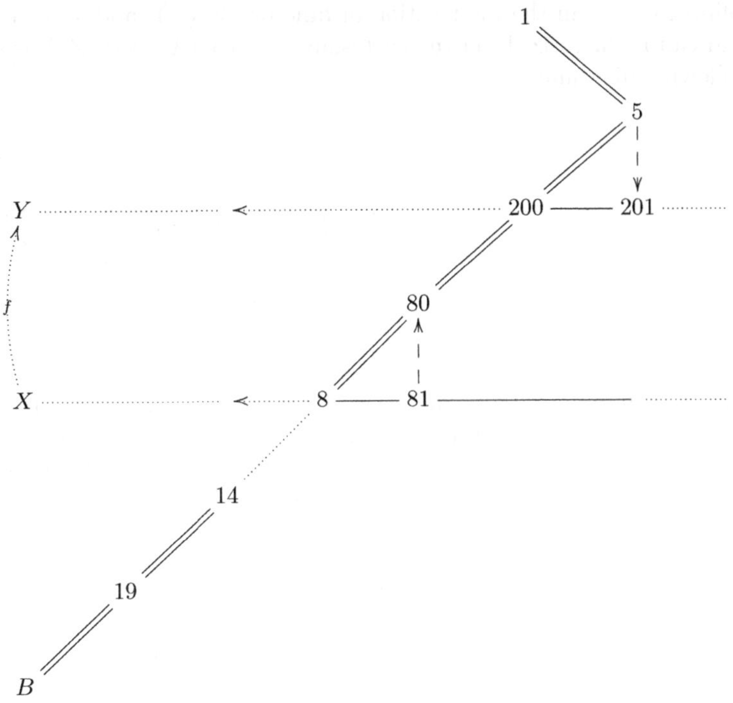

3.5 Functorial Filters

One may restrict morphisms $f : X \to Y$ to those which factor through a colimit functor defined by a diagram (X_ι). Given a functor morphism $h : \mathrm{colim}_\iota X_\iota \to @Y$, we denote the subfunctor of $@Y$ which is defined by such h-filtered morphisms $X \to \mathrm{colim}_\iota X_\iota \to Y = h \circ v$, $v : X \to \mathrm{colim}_\iota X_\iota$, by $@_{(X_\iota)}Y$. Here is a first proposition concerning such subfunctors for variable colimit filtering.

Proposition 2. *Let $@_{(X_\iota)}Y, @_{(Z_\lambda)}Y$ be two subfunctors of $@Y$ that are defined by diagrams $(X_\iota), (Z_\lambda)$ and their morphisms h_X, h_Z, respectively. Then*

$$@_{(X_\iota)}Y \cap @_{(Z_\lambda)}Y = @_{(F_{\iota,\lambda})}Y,$$

where $F_{\iota,\lambda}$ is the fiber product functor $@X_\iota \times_Y @Z_\lambda$.

Proof. In this proof we write the diagrams (X_ι), etc. instead of their colimits to ease notation. Note that we are working in the presheaf topos $\mathcal{LDC}_R^{\Sigma @}$, where limits and colimits exist, moreover colimits are stable by base change, which in particular means that the cartesian product of colimits is isomorphic to the colimit of the cartesian products of its cofactors. Let us give a morphism f with two factorizations $f = h_X \circ u : X \to \mathrm{colim}_\iota X_\iota \to Y$ and $f = h_Z \circ v :$

$X \to \mathrm{colim}_\lambda Z_\lambda \to Y$ in the intersection of functors $@_{(X_\iota)}Y$ and $@_{(Z_\lambda)}Y$. This morphism can be factorized via the cartesian product $(X_\iota) \times_Y (Z_\lambda)$ as shown in the following diagram:

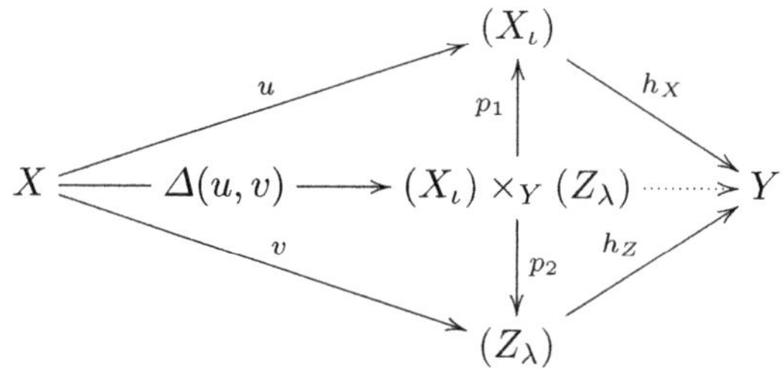

Hence the function f also factorizes by $f = h_X \circ p_1 \circ \Delta(u,v) = h_Z \circ p_2 \circ \Delta(u,v)$ through the functor $(X_\iota) \times_Y (Z_\lambda)$ with its functorial morphism $h_X \circ p_1 = h_Z \circ p_2$. But the latter is isomorphic to the colimit of the diagram $(F_{\iota,\lambda})$ of functors $F_{\iota,\lambda} = @X_\iota \times_Y @Z_\lambda$. The converse is immediate by the definition of the fiber product.

Remark 1. This proposition implies that, intuitively, multiple simultaneous content filtering can always be achieved by a single filter. Logically speaking, the conjunction of content filters is covered by a single new content filter. Trivially, the disjunction of filters also can be covered by a single filter, namely the one defined by the coproduct of two diagrams. For the time being, we don't know about the existence of a negation filter "$\neg(X_\iota)$" for a given filter (X_ι).

3.5.1 Functors for Filters

The construction of functorial filters has the following Yoneda-type background. If $\mathcal{C}$ is a category, call $\mathcal{C}^@/?$ the category of all fiber categories (which are in fact topoi) $\mathcal{C}^@/@Y$, Y in $\mathcal{C}$. We have a functor $/? : \mathcal{C} \to \mathcal{C}^@/?$ that sends a morphism $f : Y \to Z$ in $\mathcal{C}$ to the canonical functor $/f : \mathcal{C}^@/@Y \to \mathcal{C}^@/@Z$, left adjoint to the corresponding logical base change functor. The filter construction is a natural transformation $\mathit{Filter} : /? \to /?$. In fact, we have the following commutative square

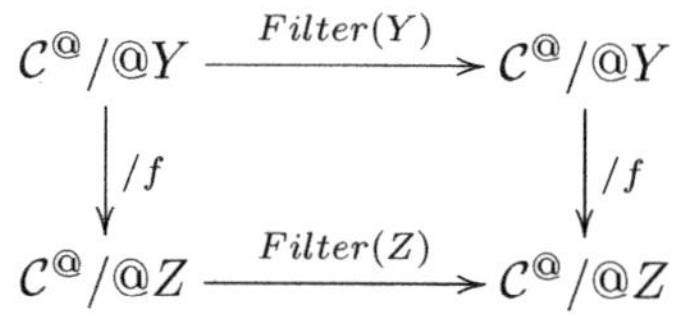

for every $f : Y \to Z$, where the functor $\mathit{Filter}(Y) : \mathcal{C}^@/@Y \to \mathcal{C}^@/@Y$ maps the fiber object $h : F \to @Y$ to the filter subcategory $@_F Y \subset @Y$. Moreover, if F is a subfunctor of $@Y$, then $\mathit{Filter}(F) = F$. The above proposition can be restated by this corollary for our semiotic context:

Corollary 1. *The natural transformation Filter* : /? → /? *transforms fiber products into intersections of subcategories, i.e., it preserves fiber products. It is the identity on the subset* $Sub(@Y)$ *of subfunctors* $F \subset @Y$ *in* $\mathcal{LDC}_R^{\Sigma @}/@Y$.

Therefore, we may consider the fiber $Filter^{-1}(G) \subset \mathcal{LDC}_R^{\Sigma @}/@Y$ of a subfunctor $G \subset @Y$. Taking $G = Filter(F)$, we call this set the *cofilter fiber* of a functor $F \in \mathcal{LDC}_R^{\Sigma @}/@Y$. The cofilter fiber comprises all functors that have the same filter effect on $@Y$, i.e., that produce the *same selection of content* Y, which is an important information about the type of ambiguity underlying semantic processes.

3.5.2 Filter Equivalence

A filter $h : F \to @Y$, which we now denote by F/Y if h is clear, gives rise to a subfunctor $@_F Y \subset @Y$. Clearly $@Y/Y$ produces $@Y$. For two filters $F/Y, G/Z$, we introduce the *filter equivalence* relation $F/Y \sim G/Z$ iff $@_F Y \xrightarrow{\sim} @_G Z$. Filter equivalence means semantically speaking that under certain filters, two H-jets play the same role as content instances. By Yoneda, $@Y/Y \sim @Z/Z$ iff $Y \xrightarrow{\sim} Z$. The general case of equivalence is less characteristic, but it conserves subfunctor relations. We write $H/Y \subset F/Y$ iff $@_H Y \subset @_F Y$.

Lemma 2. *If* $F/Y \sim G/Z$ *and* $H/Y \subset F/Y$, *then there is* $E/Z \subset G/Z$ *such that the square diagram*

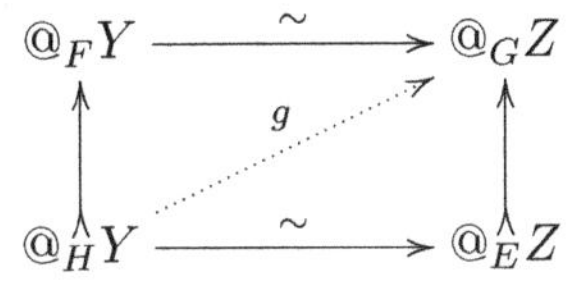

with the vertical subfunctor arrows and horizontal isomorphisms commutes.

This lemma tells us that filter equivalence conserves subfilters. Its proof is immediate: Take for E the subfunctor $E \subset @_G Z$, image of the monomorphism $g : @_H Y \rightarrowtail @_G Z$. This enables the isomorphism class subfunctor relation:

Definition 3. Denote by $[F/Y]$ the class of filters G/Z that are equivalent to F/Y. Then we say that $[F/Y]$ is a *superclass* of class $[H/X]$ *(*or that $[H/X]$ is a *subclass* of $[F/Y]$*)* iff there is $K/Y \in [H/X]$ such that $K/Y \subset F/Y$. This is denoted by $[H/X] \subset [F/Y]$.

Let us show that this property is well defined. Suppose $F/Y \sim E/T$. Then the above lemma proves that E/T has a subfilter L/T that is equivalent to K/Y, and therefore also to H/X.

3.5.3 Remarks on Full Filters

We call a filter F/Y *full* iff the corresponding functor $@_F Y$ is isomorphic to the full functor $@Y$. Concretely, we have to deal with the question when we have $X@_FY \xrightarrow{\sim} X@Y$. Take a morphism $f : X \to Y$. Suppose that all its factor arrows are sub-H-jets. Then X is a sub-H-jet of Y. If the morphism has a factorization $f = f_n \circ \ldots f_1$ where not all factors are sub-H-jets, then take the last one, f_k, of this type. Then $f_n \circ \ldots f_{k+1}$ is a sub-H-jet of Y. The morphism f_k acts on an elementary component and can be factorized by $f_k = f_k^1 \circ f_k^2$ where f_k^1 is an embedding of an elementary H-jet and f_k^2 is a morphism into that elementary H-jet. This implies that since one may evidently take F as the colimit of all sub-H-jets of Y, morphisms $X \to Y$ for non-sub-H-jets of Y can all be covered by/factorized through the elementary sub-H-jets of Y, while larger sub-H-jets of Y are needed to cover non-elementary sub-H-jets of Y only.

3.5.4 Filters and Grothendieck Topologies

Filters qua subfunctors $@_FY \subset @Y$ of representable functors are what in topos theory are called[2] *sieves* [36, I.4]. They are the basic objects of Grothendieck topologies [36, III.2]. This implies that in our context, filters could be used to define Grothendieck topologies in semiotics. For example, a given collection $\mathcal{F}$ of filters gives rise to the smallest Grothendieck topology $\langle \mathcal{F} \rangle$ containing the elements of $\mathcal{F}$. And this in turn generates the Grothendieck topos $Sh_R^\Sigma(\mathcal{F}) := Sh(\mathcal{LDC}_R^\Sigma, \langle \mathcal{F} \rangle)$.

The Grothendieck topology axioms make sense for semiotic perspectives, too. For example, axiom (ii) in [36, III.2] states that, for a member $@_FY$ of the topology, if $f \in X@_FY$ and if $h : Z \to Y$, then the fiber product $X@h^*@_FY \subset X@Z$, which is composed of those $g : X \to Z$ such that $h \circ g$ factors through F, is also in that topology. Semiotically speaking this means that we take those g, where X is an expressive instance of Z such that when Z is viewed as an expressive instance of Y via h, then X becomes an expressive instance of Y that 'lives' in F.

The main problem with such Grothendieck topologies is the criteria for selecting covering sieves. On the one hand, one does not want the discrete topology to be considered, but on the other, every filter should apriori be a potential candidate for such a topology. The selection process is a decisive topic for future research.

3.6 Functors and Trees

The very definition of H-jets is somewhat restrictive since every sign has its threefold ramification into expression, signification, and content. In our definition of H-jets we always take a selection of a linear trajectory within the multiple possibilities. In reality, such H-jets are embedded in trees with more complete ramifications such as the following configuration.

[2] Sieves are also called filters in order theory. We prefer the term "filter" in our context.

This tree is essentially the colimit of three H-jets: $H_1 = Ex(Y) \ldots \to Z$, $H_2 = Ex(X) \ldots \to Z$, and $H_3 = Ct(Sg(X)) \ldots \to Z$. H_1 and H_2 are glued together along the sub-H-jet $H_{12} = Y \to Z$, H_1 and H_3 are glued together along the same sub-H-jet H_{12}, and H_2 and H_3 are glued together along $H_{23} = X \to Y \to Z$.

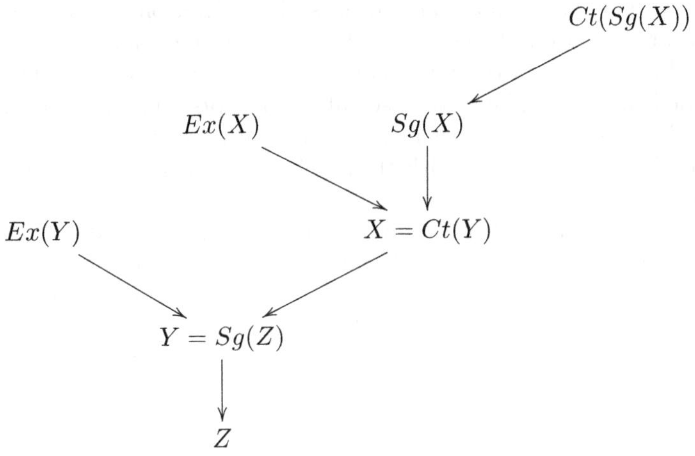

In order to describe this (by definition connected) tree T, we have to consider the functors $@H_1, @H_2, @H_3, @H_{12}$, and $@H_{23}$ and the corresponding diagram of inclusions inherited from the sub-H-jet morphisms $H_{12} \subset H_1$, $H_{12} \subset H_2$, $H_{23} \subset H_2$, $H_{23} \subset H_3$, $H_{12} \subset H_{23}$. The tree T is described by the (non-representable) colimit $C(T)$ of this diagram $\mathcal{D}$ of representable functors, it is also called a *W-tree* as it contains the H-jet $W = Y \to Z$. Also, the question of morphisms between trees of H-jets is dealt with in the functorial (Yoneda-typed) setup. Whenever we consider such tree-induced functors, we "represent" them by the defining trees, as shown above, although they are not representable functors.

In semantics, the filters defined by above tree types are very important since when a concept that is represented by a H-jet W, it is connected to the trees where it appears as a sub-H-jet. This tree set is usually extremely large and expresses the connectivity of W in the total semiotic environment. Consider now filters F/Q of a H-jet Q that are defined on domain functors that are W-tree functors, $F = C(T)$. If T_1, T_2 are two such W-trees for filters $T_1/Q, T_2/Q$ such that the morphism $C(T_1) \to Q$ factors through the morphism $C(T_2) \to Q$, we get a diagram of W-tree functors $Filt(W)Q$ of all filtered W-tree functors $@_T Q$ for a given sign or H-jet W and H-jet Q. One may then step to the colimit of all functors $@_T Q$ within $Filt(W)Q$, let us denote this functor by $Filt(W)@Q$, yielding the colimit functor morphism $Filt(W)@Q \to Q$.

Such a method seems to be enforced by widespread philosophically colored claims regarding music semiotics. For example, in [41], Viatscheslav Medshevsky, in the first page of his chapter about musical intonation, states that "musical semiotics has no other choice but to try to solve its *main problem* (my emphasis) — how "sound" and "thought" relate to each other through the basic elements of music and through such formations as "genre", "style", and the whole morphological system of a music culture...." This statement is, semiotically speaking, very broad in several respects. To begin with, thought and sound are not merely musical concepts. Therefore, on which thought or sound concept should one focus? Second, the requested relationship depends on the possible contents of these two concepts. But if one looks at them, they are embedded in a multiplicity of associated concepts, such as, for example, logic, formalism, acoustics, Fourier theory, poietical versus aesthesic aspects thereof, etc. Therefore, to deal with those relations is a very broad and fuzzy assignment.

To manage such a plethora of fuzzy conceptual and procedural neighborhoods englobing the two concepts of sound and thought, it seems reasonable to start working with the above tree functors. They describe in a precise way what a neighborhood of a concept could be. We would then consider the two filter diagrams

$$Filt(Sound)Thought \quad \text{and} \quad Filt(Thought)Sound,$$

for the two elementary H-jets $Sound = Ct(sound) \to sound, Thought = Ct(thought) \to thought$, together with the corresponding colimit fiber functors.

$$Filt(Sound)@Thought \quad \text{and} \quad Filt(Thought)@Sound.$$

The difference between the two diagrams is that in $Filt(Sound)Thought$ we look at all morphisms (perspectives) of $Thought$ that filter through trees that extend $Sound$, i.e., of structures of thought that embrace sound. This could be a conceptualization of sound that projects to the concept of a thought. The second one, $Filt(Thought)Sound$, is the totality of perspectives of $Sound$ that embrace $Thought$. This could be a conceptualization that throws light onto the understanding of sound. These formulas are a semiotic device for a future investigation of that "main problem". We are *not* claiming here that they as such solve the problem, but we claim that they should be dealt with when attempting to solve it. They are the *semiotic strategy* we propose. We have chosen this problem since it evidences the semiotic in-depth structure of any attempt to answer the question. Other difficult pairs of concepts would also be useful, such as, for example, "thinking" and "making".

We conjecture that this representation of a mutual relationship could eventually be handled by a computer software because such a huge amount of data cannot be understood directly by humans. This is perhaps the moment to think about a computer-aided semiotics that, similar to computer-aided mathematical calculations, could help humans operate on a more intense level upon semiotic details.

It is interesting to look at the difference of the two above perspectives. More generally speaking, let us look at two H-jets A, B and at the corresponding two perspectives, i.e., $Filt(A)B, Filt(B)A$ together with the colimits $Filt(A)@B, Filt(B)@A$. Suppose that $Filt(A)B$ is not empty, which implies that there is a tree T for A and a morphism $@_T@B$. This means that there is a morphism from each cofactor H-jet U in T to B. But this implies that A is connected to B, i.e., every A-tree is contained in a B-tree. But also conversely: Every B-tree is contained in an A-tree. *This means that the ordered set of A-trees has a cofinal subset that consist of trees $T_{A\&B}$ that are A- and B-trees. The same holds for B-trees.* Therefore, we may limit the colimit construction to such symmetrical A- and B-trees. We have

$$Filt(A)@B = \text{colim}_{T_{A\&B}}@B \quad \text{and} \quad Filt(B)@A = \text{colim}_{T_{A\&B}}@A$$

However, this does not imply that the functors $Filt(A)@B, Filt(B)@A$ are in any way symmetrical. It may very well happen that, while $Filt(A)@B$ is not empty, the other one, $Filt(B)@A$ is empty. The existence of morphisms is not a symmetrical fact, only a cofinal tree set is so. But the colimit functor $\text{colim}_{@T_{A\&B}}$ is the maximal semiotic "neighborhood" of A and B that could matter for any semiotic relationship between A and B. One should however be aware that in general, $\text{colim}_{T_{A\&B}}@B \subsetneqq @B$. In fact, if we are given an element $f : X \to B$ of $X@B$, then X is in $T_{A\&B}$, and f can be factorized via $f = f \circ Id_X$, but f cannot, in general be extended to a morphism $g : \text{colim}_{T_{A\&B}} \to B$.

3.7 The Creative Evolution of Semiotic Categories: Time in Categories?

In category theory, time-dependence of a category is not conceived. But in our situation, the repertory of signs might be a function of time since new signs or concepts may arise as our human knowledge expands. This is an essential topic regarding the question of creative extension of a semiotic system [32]. The core extension is not really a time problem, but the problem of a structural extension, similar to algebraic extensions of fields in algebra. Similar to that algebraic concept architecture, we are now facing the creativity process and hoping that it can be described in a maximally constructive way.

The process of (musical) creativity is composed of the following steps [32, p. 17]:

1. Exhibiting the open question
2. Identifying the semiotic context
3. Finding the question's critical sign or concept in the semiotic context
4. Identifying the concept's walls
5. Opening the walls and displaying its new perspectives
6. Evaluating the extended walls

Let us see, how much of our present functorial semiotics can help formalize this scheme into mathematics. The initial step (1) is kept as is, while the second step (2) is rightly interpreted in terms of our category $\mathcal{LDC}_R^\Sigma$. Step (3) boils down to finding a H-jet Y that plays the role of the critical sign, this is not problematic either.

While step (6) is straightforward, steps (4) and (5) are more delicate. In the algebraic setup, one would take the defining polynomial $P(X)$ as the critical concept, while the question would be about the (non)existence of P's roots in the given field K. The construction of roots is following a very interesting "softening of walls", namely the passage to the polynomial algebra $K[X]$, where P exists without any allusion to roots. The extension of K to a field L that contains roots of P is then achieved by the passage of $K[X]$ to the quotient ring $L = K[X]/(P)$.

This is a model we should follow in semiotics, too. And in such a way that the algebraic process is a *special case* of the general semiotic one. We shall discuss this semiotic methodology in details in Chapter 10.

4

Examples

Summary. This chapter presents a number of essential examples of H-jets: pointers, sets,

– Σ –

4.1 A Classical Example from Music

A classical example of H-jets in music is the connotational chain from the score text to emotions that are generated when listening to a performance, see also [34, p. 83]. Here is the H-jet architecture for this process:

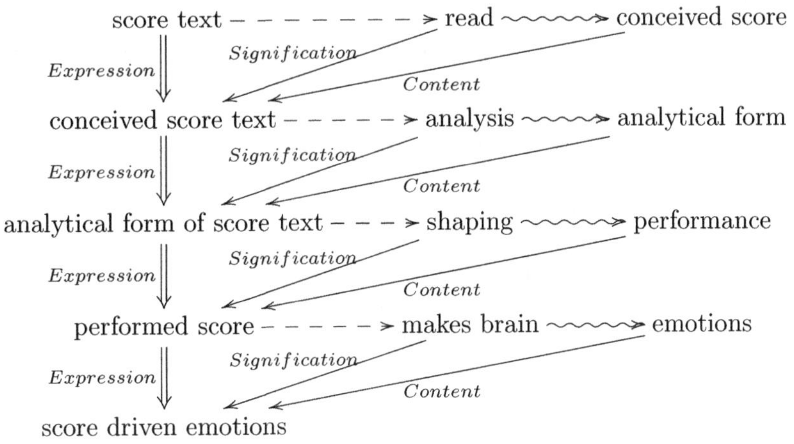

Opposed to this H-jet is the improvisation-based evocation of emotions. It could be represented as follows:

G. Mazzola et al., *Functorial Semiotics for Creativity in Music and Mathematics*, Computational Music Science, https://doi.org/10.1007/978-3-030-85190-3_4

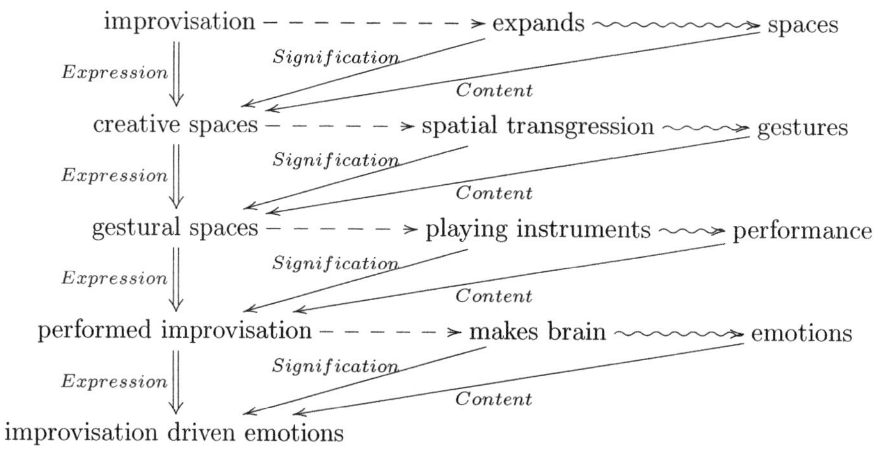

Here we have the left chain of expressions and double arrows (=>) to be taken as a H-jet. The other components are only shown to illustrate the potential of H-jets that can be constructed from this configuration. This example makes also evident that the structure of a H-jet with its components is relevant to semiotics. Only looking at emotions (lowest level here) is a completely different semiotic situation than looking at the entire trajectory from the score text or improvisatory expression to those emotions.

4.2 Pointers

It is well known [42, 43] that apes don't point, and that this basic action is learned by humans at the age of one year, preceding and very probably also enabling the development of human language, which is the prototype of a semiotic system. In the core momentum of conceptualization of signs, it is essential to recognize that the very structure $X = Ex(X) \dashrightarrow Sg(X) \rightsquigarrow Ct(X)$ of a sign X is based upon the direction from $Ex(X)$ to $Ct(X)$ that is given by the signification component $Sg(X)$ which stands for the Latin "pro" vector. This directionality is not a semiotic, but a *presemiotic* concept, signs can only be defined if the vector of signification is supposed to act. We have argued for this insight in different publications in the sense that the 'movement' of signification is essentially a *gesture* proper, not a sign.

To deal with this gestural condition on a sign's X signification component $Sg(X)$, we now focus on the case where this component may be a mathematical gesture, see [31] for a reference to the theory of mathematical gestures. In this setup, where we say that *a sign has a pointing signification.* Such a signification component is defined to be a mathematical gesture $\gamma : \Gamma \to \overrightarrow{\mathcal{LDC}}$, which means that $\mathcal{LDC}$ is viewed as an indiscrete topological category. This implies that γ can be any digraph morphism from digraph Γ to the digraph of curves $\overrightarrow{\mathcal{LDC}} = \nabla@\mathcal{LDC}$ in the category $\mathcal{LDC}$.

Let us now look at a first example of a sign P

$$\begin{array}{ccc} Ex(P) \dashrightarrow & Sg(P) & \rightsquigarrow Ct(P) \\ \diagdown & | & \diagup \\ & P & \end{array}$$

that is essentially a pointer. We set $Ex(P) = PT$, PT being a character string (a word), the pointer's name. The content $Ct(P)$ is the irreducible concept of a *pointer gesture*, which we denote by "POINTER". The signification component is a simple mathematical gesture $\gamma_P : \uparrow \mapsto f_P$, defined on the arrow digraph $\uparrow$ that has two vertices and one connecting arrow. Here f_P is the curve[1] $\searrow (E \circ F)$ of the composition $E \circ F$ of the two elementary morphisms $F : (Ex(P), P) \rightarrow (Sg(P), P)$ and $E : (Sg(P), P) \rightarrow (Ct(P), P)$. Observe that P is defined in a non-vicious circular way as its signification component refers to P. This sign is essentially a pointer, nothing more, but it is crucial in the development of more complex signs. We may use $Sg(P)$ as a minimal pointing signification component for other signs. So the content of P, a veritable gesture, is signified through the mathematical gesture concept from mathematical gesture theory. It is important that a gesture can be given as a content, but not as a sign since gestures are presemiotic.

4.3 ZF Set Theory

The very first fundamental corpus of signs should be those that mathematics are often using as a default basis: sets. We propose the following approach to ZF (Zermelo-Fraenkel) set theory. The first action we need is to specify sets among the totality of general signs. We do this by defining sets as signs of the following shape:

$$\begin{array}{ccc} Ex(X) \dashrightarrow & \gamma_P & \rightsquigarrow SET \\ \diagdown & | & \diagup \\ & X & \end{array}$$

When this is given, we need to introduce the element relation $x \in X$ among sets. We shall do this specifying signs which have this architecture:

$$\begin{array}{ccc} X \dashrightarrow & E & \rightsquigarrow x \\ \diagdown & | & \diagup \\ & X\beta x & \end{array}$$

The signification E is defined to be the sign

[1] This is defined in [31].

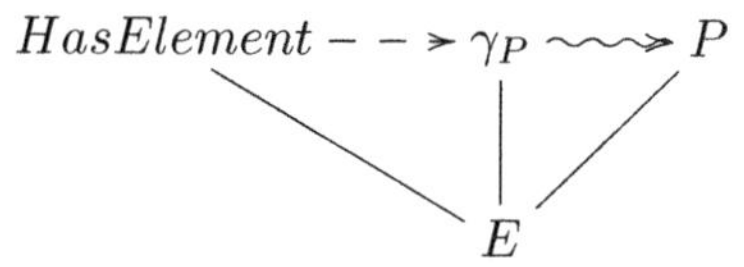

which is a synonym[2] of the pointer P. Of course, the sign $X\beta x$ does not determine X, but it is a contribution to the content of X. After all, what is the meaning of a set? It's its elements. This is a first step towards global contents. A set is a homonymous sign that is identified (as a set!) by all of its element-relational signs. We may now define subsets by the usual $X \subset Y$ iff $X\beta x$ implies $Y\beta x$ for all sets x. Equality then means $X \subset Y$ and $Y \subset X$, and the axiom of equality "$X\beta x$ and $x = y$ implies $X\beta y$" initiates the usual ZF set theory.

Remark 2. There is a special semiotic situation here: When we write $X\beta x$, and $X\beta y$, this means that X is an expression of two contents, a homonymous setup. However, X is not identified as being a H-jet here, so there is no diagram of morphisms $x \leftarrow X \rightarrow y$. This is due to the fundamental setting in our category, namely that morphisms must involve H-jets as objects. X is not such an object. This is correct in that only semiotic relationships can provide us with an object status. The elementary H-jets X—$X\beta x$ and X—$X\beta y$ are different, therefore X is not embedded in a connecting object for this homonymy. We propose to solve this formal problem by introducing a very reduced sign, call it ∇X, for X, namely

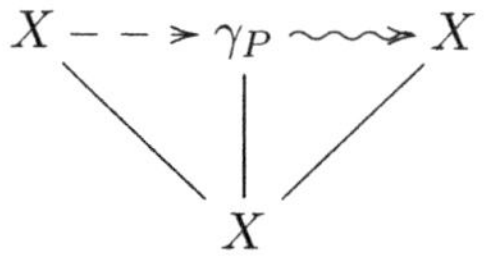

This construction transforms X into a H-jet, and we may consider this diagram:

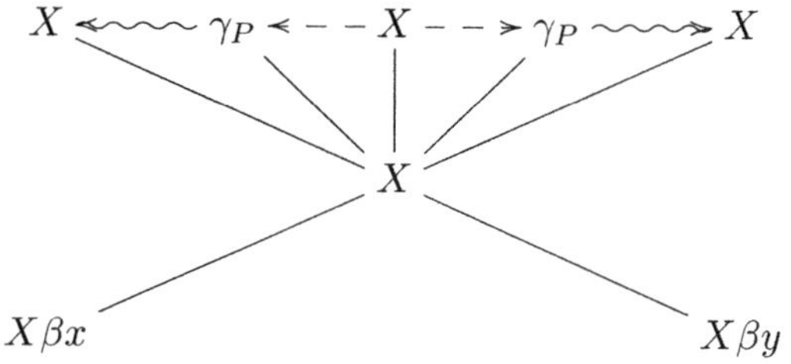

which now connects the elementary H-jet X—X by two morphisms to the H-jets X—X—$X\beta x$ and X—X—$X\beta y$. We should view every such situation as a reduced representation of the ∇X sign. With this convention, X synonymy/homonymy can be viewed as an instance of $@\nabla X/\nabla X@$.

Lemma 3. *If $f : a \rightarrow b$ is a function and if $x \in a \cup b$, then there is a morphism $p : f \rightarrow x$ in $\mathcal{LDC}$.*

[2] The symbol β was introduced by Paul Finsler in his non-formal set theory [10].

The proof is trivial once we view the β set relation as a morphism and recall that $f : a \to b$ means the triple (f, a, b), which means $((f, a), b)$, where an ordered pair (u, v) of sets means $\{\{u\}, \{u, v\}\}$, and we are done using the β relation morphism. More precisely, if we are given an element relation

$$\begin{array}{ccc} x \dashrightarrow & E & \rightsquigarrow y \\ \searrow & \downarrow & \swarrow \\ & x\beta y & \end{array}$$

and a second element relation

$$\begin{array}{ccc} y \dashrightarrow & E & \rightsquigarrow z \\ \searrow & \downarrow & \swarrow \\ & y\beta z & \end{array}$$

we can view these relations as morphims $u : (x, x\beta y) \to (y, x\beta y)$ and $v : (y, y\beta z) \to (z, y\beta z)$. Using Lemma 3, we can extend the codomain $c_{\cdot} = (y, x\beta y)$ of u and the domain $d_{\cdot} = (y, y\beta z)$ of v to $c^*_{\cdot}$ and $d^*_{\cdot}$ to get a morphism $w : c^*_{\cdot} \to d^*_{\cdot}$. Prepending the extension $c_{\cdot} \to c^*_{\cdot}$ to w and extending v to the larger domain $d^*_{\cdot}$ in the obvious way, we get the desired morphism $(x, x\beta y) \to (z, y\beta z)$.

Corollary 2. *For any set X, there are $\mathcal{LDC}$ morphisms $(\gamma_P, X) \to (Ex(P), P)$, $(\gamma_P, X) \to (Sg(P), P)$, and $(\gamma_P, X) \to (Ct(P), P)$.*

To understand this, observe that γ_P is essentially a function, while the codomains here show up as elements in the function's codomain.

4.4 A Second Example from Music

Let us look at the H-jet Y_{score} of score performance (starting with a *score text*) as described in Section 3.1.1. Let us look at the functor F that is the coproduct of two special elementary sub-H-jets X_1, X_2 of Y_{score} given by

$$X_1 = \begin{array}{c} \text{score text} \\ \Downarrow \\ \text{conceived score text} \end{array} \qquad X_2 = \begin{array}{c} \text{performed score} \\ \Downarrow \\ \text{score driven emotions} \end{array}$$

To define a functor above Y_{score} (what will be called a filter in Section 4.6.3) F/Y_{score}, take the morphism $F \to Y_{score}$ given by the two inclusions $X_1 \to Y_{score}, X_2 \to Y_{score}$. Then the set $X@_F Y_{score}$ of the morphisms $X \to Y_{score}$ that factor through F describes morphisms of domain X that focus on (factor through) two particular aspects of the score performance H-jet Y_{score}, either the elementary H-jet X_1, or X_2, either the transition from score text to conceived score text, or the transition from the performed score to the score driven emotions. This mathematical construction formalizes the musicological alternative between the two analytical perspectives X_1 and X_2.

4.5 The Role of Signification

A major topic of functorial semiotics is the role of a sign's X signification $Sg(X)$ in the production of its content $Ct(X)$. From Lemma 3 it follows that all elements, qua sets, intervening in the signification $Sg(X) = \gamma_X$ of X, are reached from $Sg(X)$ as codomains of morphisms of $\mathcal{LDC}^{\Sigma}_{R}$. We also should recall that such elements might be signs of another domain, having contents of different, non-set-theoretic, nature. This can *a fortiori* happen to $Ct(X)$.

With these facts in mind, the question arises about how the content $Ct(X)$ can be deduced or produced from this $Sg(X)$ information that one can view as being generated by the covariant functor $Sg(X)@$. In the limit, one would expect that a type of algorithm could generate $Ct(X)$ from $Sg(X)$. The question arises, where such a processual machine could reside. The very role of signification suggests that this should also occur within $Sg(X)$. This seems to be more complex ontologically than a superficial view would make one believe. In fact, if a process, such as an algorithm, is to be instantiated to reach $Ct(X)$, who would be the subject operating such a process? Should one include this activity within $Sg(X)$, and how must this be realized? A priori, there is no semiotic reason why the content of a sign could not be an activity that is performed by a specific subject. But the technical realization of such a process must be made precise.

4.5.1 An Elementary First Example of Signification

The following is an analysis of this significative process. To understand the issue of signification processes, let us have a closer look at a very simple signification process in arithmetics, of natural numbers, say. One defines the sum $a + b$ of two numbers a, b and then defines the sum $a + b + c$ of three numbers by $a + b + c := (a + b) + c$. Semiotically speaking, this looks as follows. Let us consider the overall sign. We have an expression "$a + b + c$" that signifies the sum $(a + b) + c$. Let us call this sign $Sum(a, b, c)$, getting

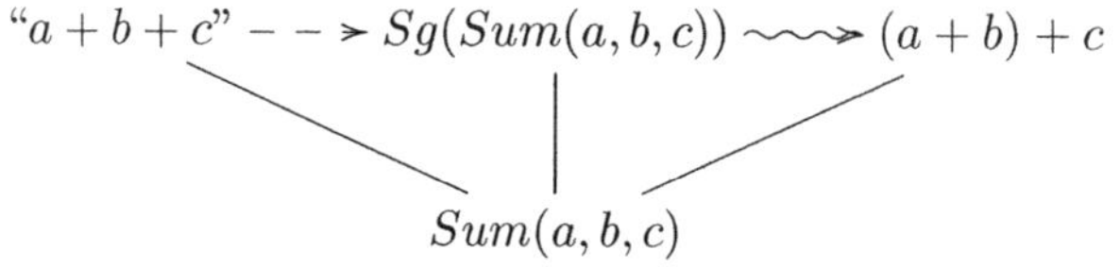

The signification component looks as follows. We first have to get the meaning of "$a+b+c$", which is "("$a+b$")$+c$", yielding the content of the part within the brackets, the sum of a and b, and therefore "$(a + b) + c$". This in turn yields its content, the sum $(a+b)+c$ of the two numbers $a+b$ and c. This adds up to the configuration

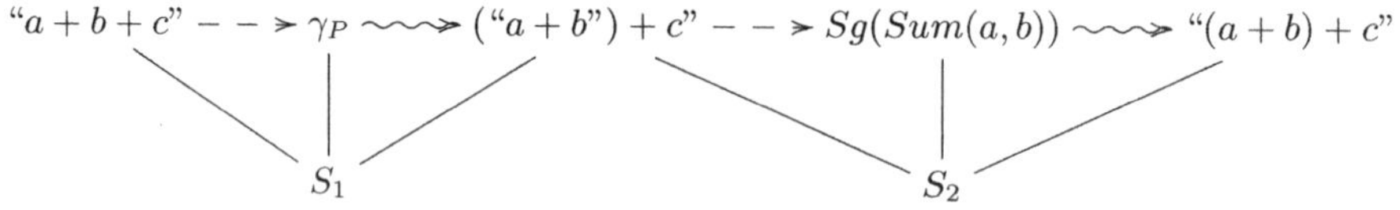

that is continued by

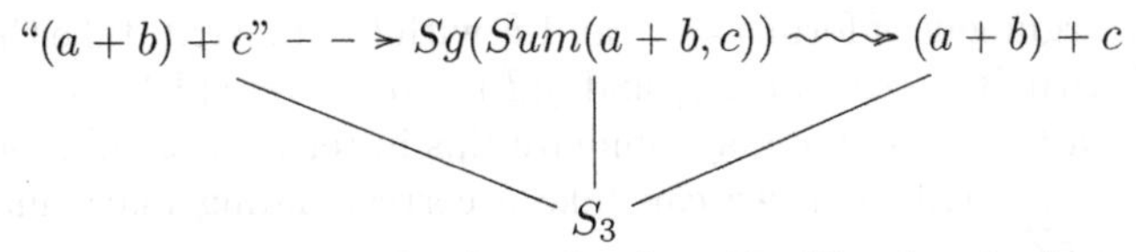

with the three sign symbols S_1, S_2, S_3. The first signification is the arbitrary case γ_P, while the second and third significations are the processes that define the sum of two numbers, and which we have not made explicit here. We can therefore summarize the signification process of $Sum(a, b, c)$ as the morphism f in $\mathcal{LDC}$ that results from this configuration:

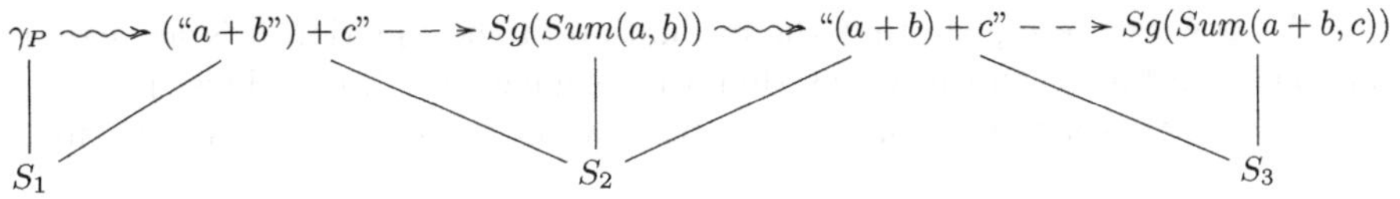

Remark 3. We may formally define this signification component by a mathematical gesture $\gamma_{Sum(a,b,c)} : \Gamma \to \overrightarrow{\mathcal{LDC}}$ that specifies the four morphisms that add up to f, taking $\Gamma = \cdot \to \cdot \to \cdot \to \cdot \to \cdot$ and thereby defining the sign

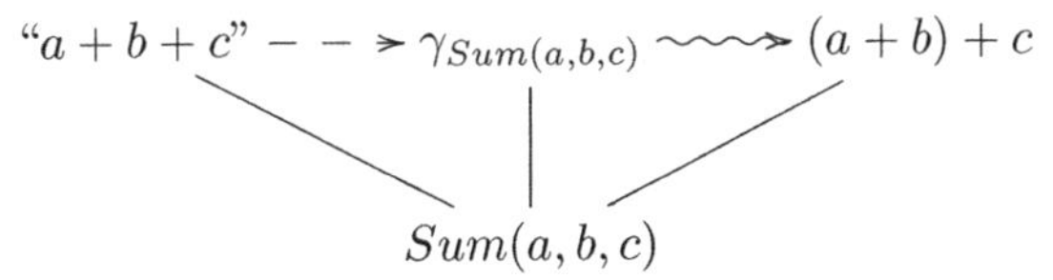

of the triple sum.

While in this example the first signification component, the arbitrary γ_P, is completely algorithmic, the significations $Sg(Sum(a, b))$ and $Sg(Sum(a + b, c))$ are less evident. It looks like they are two instances of a more general signification of sums, once applied to a, b, then to $a + b, c$.

4.5.2 The Case of Logical Signs

When it comes to design signs that embody mathematical definitions such as the sum of natural numbers a, b, it immediately turns out to be a conditio sine qua non to embrace logical decisions, too. In the case of the sum $a + b$ of natural numbers, for example, we have to distinguish two cases: either $b = 0$ or $b = c^+$, the successor of c. In the first case, we set $a + 0 = a$, else, we set $a + c^+ = (a + c)^+$. We therefore need to make that decision upon b before defining the sum. Mathematically speaking, a logical function is a function $f : X \to TF = \{T, F\}$, the codomain being the set whose elements are Truth (T) and Falsity (F). According to our prior discussion (Lemma 3), such functions can be understood as being signs such that any relationship of a function to its set-theoretical elements can be stated as a sign whose content is such an element. Therefore the values of logical functions are accessible as contents of determined signs.

Suppose that we are given the logical function $f : \mathbb{N} \to TF$ that associates T with an argument of form $b = c^+$ and F with $b = 0$. Next, take the function $g : TF \to B$ with $B = \{(a+c)^+, a\}$ and $g(T) = (a+c)^+, g(F) = a$. Consider their composition $h = g \circ f$. We may simulate this in terms of $\mathcal{LDC}$ morphisms. And using this representation, we can take a corresponding mathematical gesture $\gamma_h : \Gamma \to \overrightarrow{\mathcal{LDC}}$ to define a sign

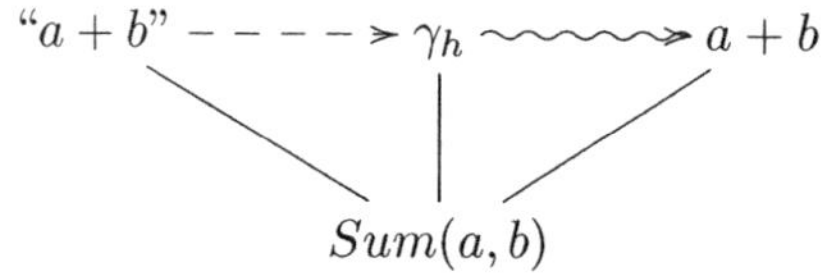

We however recognize that the evaluation of the function h is an action that must be taken, and it seems that this engages a subject. This question requires a more profound ontological analysis, which we defer for the time being.

4.6 Forms and Denotators as Signs

In mathematical music theory, forms and denotators define the basic concept architecture. We want to reinterpret them in terms of sign configurations. See also our previous semiotic interpretation of forms and denotators [26]. It might be helpful to understand this architecture as being similar to the object-oriented programming (OOP) paradigm. OOP classes correspond to forms, while OOP objects correspond to denotators, or, geometrically speaking (and in perfect congruence with Kant's saying that concepts are points in conceptual spaces): forms are spaces of a certain type, while denotators are points in such spaces.

4.6.1 Forms

A form as defined in [27, Section 6.6.2] is a quadruple $F = (NF, TF, CF, IF)$ that consists of its name NF, a character string, its type TF, which is Simple, Power, Limit, or Colimit, its coordinator CF, which is a diagram of forms or a module, and the identifier IF which is a monomorphism $Fun(F) \rightarrowtail Frame(F)$. A simple example would be the String_Quartet form, which is the colimit (disjoint union) of four instrumental forms: Violin_1, Violin_2, Viola, Violoncello, each of them being Power forms (powersets) of the Note form, which is the limit of the note parameter forms Onset (real numbers), Duration (real numbers), Pitch (rational numbers), Loudness (character strings such as *mf*), Glissando (rational numbers), Crescendo (character strings). The Power forms here also are restricted to finite sets of notes, which defines the identifier monomorphism.

To rephrase this data in terms of signs, we first define the form's type as follows: For a Simple type with module M, we set

$$\begin{array}{ccc} Simple & \dashrightarrow M \rightsquigarrow & @M \\ & SimpleType(M) & \end{array}$$

For Power type of form F we set

$$\begin{array}{ccc} Power & \dashrightarrow \mathcal{D}(F) \rightsquigarrow & \Omega^{Fun(F)} \\ & PowerType(F) & \end{array}$$

where $\mathcal{D}(F)$ is the trivial one-vertex diagram $\bullet \to F$. For Limit type we set

$$\begin{array}{ccc} Limit & \dashrightarrow \mathcal{D} \rightsquigarrow & \mathbf{Limit}Frame(\mathcal{D}) \\ & LimitType(F) & \end{array}$$

where $\mathcal{D}$ is the type's diagram of coordinator forms. And for Colimit type we set

$$\begin{array}{ccc} Colimit & \dashrightarrow \mathcal{D} \rightsquigarrow & \mathbf{Colimit}Frame(\mathcal{D}) \\ & ColimitType(F) & \end{array}$$

With these sign construction, one may define a form to be the following sign type, depending ton the type sign $MyType$:

$$\begin{array}{ccc} FormName & \dashrightarrow MyType \rightsquigarrow & IF \\ & Form & \end{array}$$

where $IF : Fun(F) \rightarrowtail Ct(MyType)$ is the form's identifier monomorphism into the type's content functor.

4.6.2 Denotators

A denotator D as defined in [27, Section 6.3.1] is a triple $D = (ND, FD, CD)$ that consists of a character string ND, the denotator's name, its form FD, and CD, an element of $M@Fun(FD)$, the denotator's coordinates at address module M. For our above String_Quartet form, a denotator would be a set of notes for Violin_2, say.

We may define the sign associated with D by this configuration:

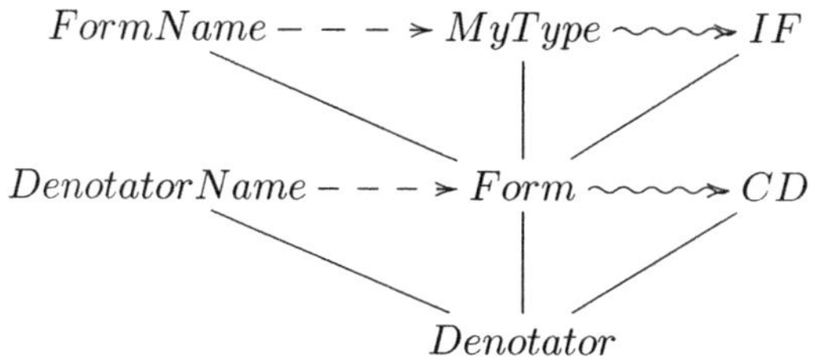

Observe that the identifier IF as well as the denotator's coordinates CD are somewhat arbitrary, i.e., they are not totally determined by the two signifiers *MyType* and *Form*, respectively. For IF, only the frame functor is given, not the subobject, and for CD, the functor $Fun(Form)$, but neither the address M nor the selected element in $M@Fun(Form)$ are given.

4.6.2.1 A Musical Denotator Example: Cadence

Let us make a first musical example, the cadence in measures 128-130 in the Allegro movement of Beethoven's "Hammerklavier" Sonata op. 106, see Figure 4.1.

Fig. 4.1: The $E\flat$ major cadence in measures 128-130 in the Allegro movement of Beethoven's "Hammerklavier" Sonata op. 106.

The cadence is this semiotic configuration of H-jets:

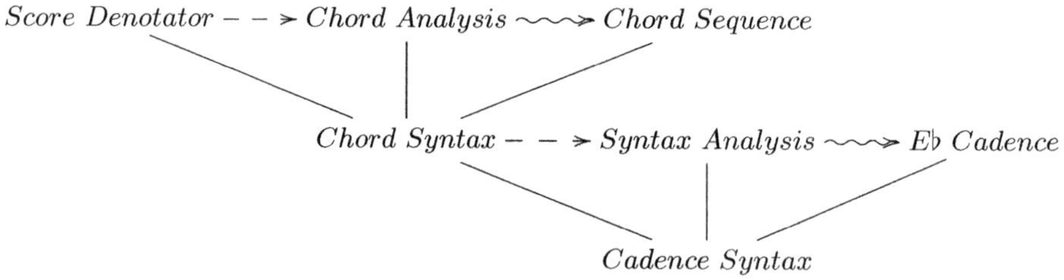

Let us explain its components. The expression of the sign *Chord Syntax* is a denotator *Score Denotator* describing the score in Figure 4.1. Its form is named "Piano NoteS", it is a *Power* form whose coordinator CF is the *Limit*

form *Piano_Note*, whose coordinate spaces are the simple forms *Onset, Pitch, Loudness, Duration* describing notes for the piano.

The motivational component *Sg*(*Chord Syntax*)= *Chord Analysis* is a sign, which performs this analysis on *Score Denotator.* It creates the content *Ct*(*Chord Syntax*)= *Chord Sequence.* The latter is a denotator of form *Chord Sequence*, which is a *Limit* of copies of the form *Chord*, which is a *Limit* of forms *Onset* and *Pitch Class Set*, the latter being a *Power* form of the simple form *Pitch Class* representing pitch classes in $\mathbb{Z}_{12}$.

The sign *Chord Syntax* is the expressive part of a connotational sign *Cadence Syntax.* Its signification process is given in the part *Sg*(*Cadence Syntax*)=*Syntax Analysis.* It results in the content *E♭ Cadence*, which is a denotator of form *Cadence.* The latter is a *Limit* form of copies of the form *Degree*, which is a simple form of character sequences (words), the words used for degrees, such as $V^7_{E\flat}$.

The *E♭* cadence $(V^7_{E\flat}, I_{E\flat})$ here is drawn from the chord sequence (in degree notation for the sake of simplicity)

$$(V^7_{E\flat}, V^7_{E\flat}, I_{E\flat}, V_{E\flat}, I_{E\flat}, V^7_{E\flat}, V^7_{E\flat}, I_{E\flat}, VII_{E\flat}, I_{E\flat}, I_{E\flat}).$$

4.6.3 Examples of Non-representable Functors and Their Semiotics

To begin with, we list three H-jets from poetology: The sign "L'Albatros" from Charles Baudelaire's synonymous poem in his collection "Les fleurs du mal" (1857), the sign "Marthe", François Villon's acrostic of his lover Marthe in Villon's "Grand testament" (1489), and the sign "Bäume Blüth" in Hölderlin's last poem "Die Aussicht" (1843). Observe that these example also illustrate the classifying role of more complex H-jets, not just elementary H-jets.

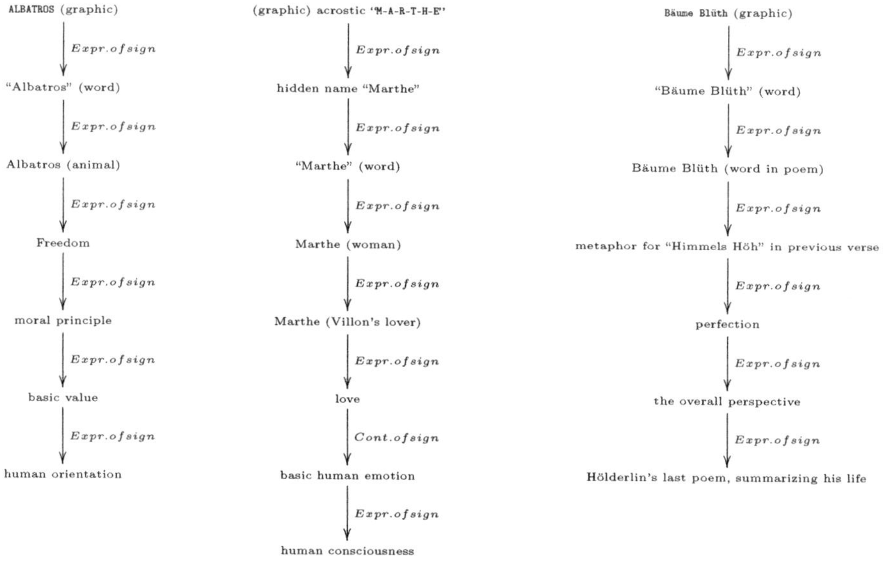

It is a good semiotic exercise to think of these three H-jets as playing the role of H-jets A, C, E in the previous general diagram, and to conceive the missing H-jets B, D, F to fill them in according to that general diagram.[3]

Let us now also describe a candidate for the composed morphism $f : X \to \mathrm{colim}_\iota X_\iota \to Y$ for the factorization through H-jet C. This generates a connection of H-jet X that expands on survival and H-jet Y that deals with a unifying human force, and which are mediated by Villon's Marthe acrostic (H-jet C in our scheme). We don't specify the ingredients of X and Y any further, the horizontal arrows are just generic placeholders for H-jet arrows that include the shown vertices. This example illustrates in a remarkably clear way the explicatory power of the category $\mathcal{LDC}^\Sigma_R$ in its connectvity that is enabled when constructing morphism from H-jets instead of simple signs. Although

[3] For example, think of B as a H-jet thematizing freedom in love as a human condition, D as a H-jet thematizing love as an attempt to perfection, and F as a H-jet thematizing the eternal perspective and God's universal plan.

this language is extremely compact, it opens up a wide semiotic discourse.

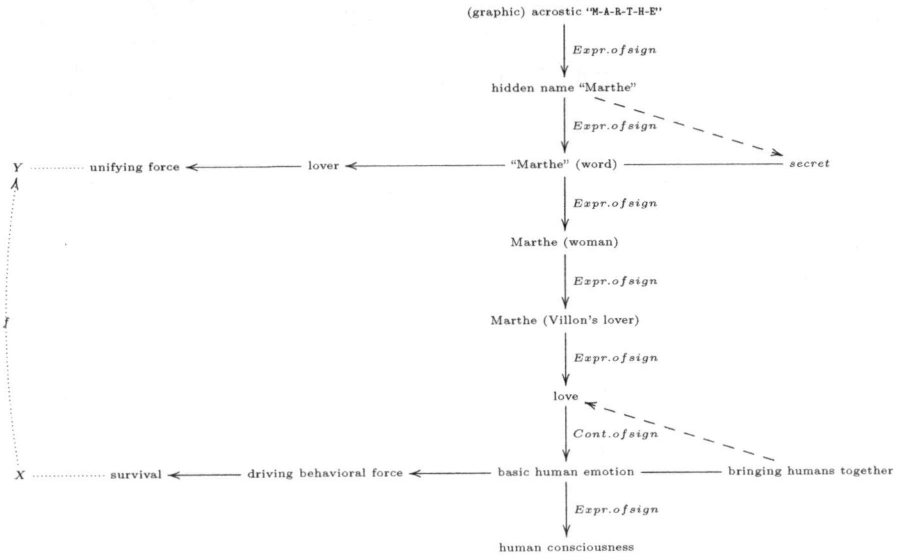

The general method of functorial factorization $X \to \mathrm{colim}_\iota X_\iota \to Y$ is useful to create *filters of content* in the sense that we are not only looking at the functor $@Y$ from address X, but we ask for morphisms $X \to Y$ which are mediated by a diagram (X_ι) of H-jets, in symbols: $X - (X_\iota) \to Y$. This means that such content instances must be channeled through specific H-jets and their morphism relations. In our previous example, unifying force as a content of survival strategies is viewed through the *filter* of Villon's love for Marthe. In common language this would say that the meaning of survival is given the perspective of that individual situation of two loving humans when reifying a unifying force. This type of concept architecture is a substantial enrichment of standard semiotics. It gives a formal representation of what it means to view a content under a *specific point of view*, given by the filter functor. Moreover, it exemplifies what we have described as a globalized content approach.

4.6.4 A Musical Content Filter: Catastrophe Modulation in Beethoven's Op. 106

This example of a content filter $X - (X_\iota) \to Y$ relates to the modulation $E\flat \to D/b$ in the Allegro movement of Beethoven's "Hammerklavier" Sonata Op. 106, measures 189-197, see the score in Figure 4.2.

Fig. 4.2: The catastrophe modulation $E\flat \to D/b$ in measures 189-197 in the Allegro movement of Beethoven's "Hammerklavier" Sonata Op. 106.

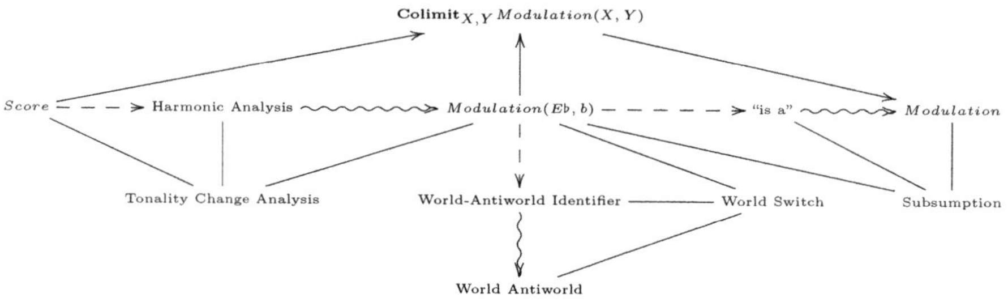

This situation is known as what Jürgen Uhde and Erwin Ratz call a "catastrophe modulation" [27, Ch. 28.2.2]. Its anatomy is radically different from "normal" modulations in this movement. This classification of modulations follows Ratz's idea that Beethoven has constructed a landscape of tonalities that splits into a "world" around the main tonality $B\flat$ major and an "antiworld" around b minor. Catastrophe modulations take place whenever the music switches from the world to the antiworld or vice versa. The present example should illustrate this situation. This means that we start with the score denotator *Score* to the left of the above diagram, the score is shown in Figure 4.2. We then cannot give it the unrestricted content of a modulation, the instance *Modulation* to the far right. Instead we apply a filter, which is the colimit of the (presheaf of) modulations $X \to Y$ that switch between the two worlds, i.e., either X is in the world and Y is in the antiworld, or vice versa. The precise set of the world and antiworld tonalities has been exhibited in [27, Ch. 28.2.3]: The world comprises the major tonalities $A, E, G, D\flat, E\flat, B\flat, C, G\flat$, the antiworld comprises $D, B, A\flat, F$, where D stands for the aeolian b mode. The colimit in our diagram ranges over all tonality pairs (X,Y) for catastrophe modulations $X \to Y$, i.e., world changes. We now filter the result of a harmonic analysis of the score through this colimit, and this is the case if the modulation that results from the harmonic analysis is a catastrophe modulation, as in our case $E\flat \to D/b$, which we represent in the diagram. This means that only catastrophe modulations make it to the modulation content here, we have installed a 'catastrophe filter'.

4.6.5 Colimits, Filters and Beethoven's Op. 109

In [32, Ch. 19.2], a mathematical model of creativity was described that was based upon the Yoneda Lemma's statement that to understand an object X in a category $\mathcal{C}$, it is sufficient to look at its functor $@X$, i.e., at all the morphisms $f : Y \to X$. But this is mostly too much, it often suffices to select a small subssystem $(Y_i)_i$ of domains together with some connecting morphisms to identify X. For example, in set theory, the singleton system $Y_0 = \{0\}$ is sufficient to identify sets up to bijections. In that model of creativity, Mazzola identifies the creative action by the exhibition of such a subssystem $(Y_i)_i$. If the category has colimits, the systems of morphisms $(f_i : Y_i \to X)_i$ corresponds to to single morphisms $\text{colim}(f_i) : \text{colim}Y_i \to X$. In terms of filters, this means that to understand X it is sufficient to look at the morphisms that factor through one of those Y_i.

This might look quite abstract, but the third movement of Beethoven's Sonata Op. 109 yields an excellent example, this has also been discussed in [32, Ch. 26]. The following example is quite compact, but we hope it can open to the reader a further trajectory from the present already long text. Beethoven creates five variations of this movement's theme and then, after all these variations have shed light (contrapuntal, rhythmical, melodic, etc.) onto the theme, he adds a sixth variation, with that very long and famous trill between b and $c\#$. Prominent musicologists, such as Jürgen Uhde and William Kinderman, have described this variation as "a streamland with bridges", which is exactly what a colimit represents: a union of parts, together with 'bridges', i.e., identification of corresponding points. The sixth variation stands for the colimit of the first five variations.

4.7 Artificial Conceptual Frameworks

Artificial Neural Networks (ANN) are the classical formal devices for deep learning approaches. In this section, we want to generalize these formalisms to concepts instead of real numbers, which are the basic arguments in ANNs. To begin with, recall that we view sets as being special categories, their elements being the categorical objects, with identities only, and functors between sets being set functions.

4.7.1 ANNs

An artificial neuron $N = (a, o, h)$ is defined by a diagram as follows:

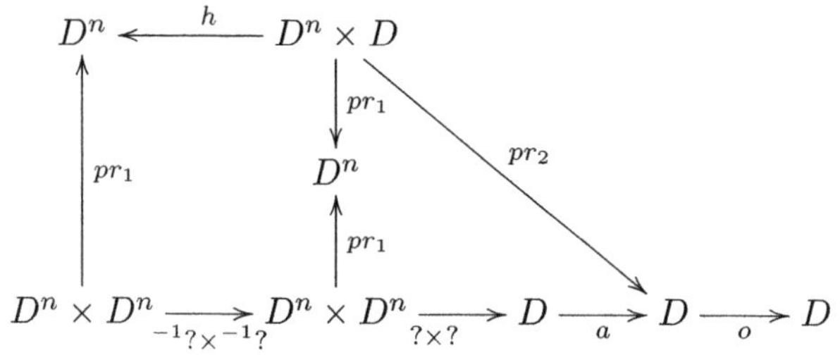

In [30, Vol. II], one associates with N its state space S_N, which is the limit of the diagram of N. One exhibits the input s_N is the n-tuple $D^n = \mathbb{R}^{n \times \mathbb{Z}}$, while the output is the D element to the extreme right of N. An ANN network is a diagram of artificial neurons, where the inputs of members of ANN are outputs of other artificial neurons, see Figure 4.3. This approach is basically numerical because all values are taken from the real numbers, elements of $\mathbb{R}$, and of cartesian powers thereof.

In our conceptual generalization, we take more generally categories $\mathcal{C}$ instead of $\mathbb{R}$ and its powers.

4.7.2 Artificial Conceptual Networks, ACNs

The idea behind our generalization is to enable values from a category $\mathcal{C}$ instead of $\mathbb{R}$. The corresponding *Artificial Conceptron* (AC) is defined by a diagram of the following type, where $D_\mathcal{C} = \mathcal{C}^\mathbb{Z}$:

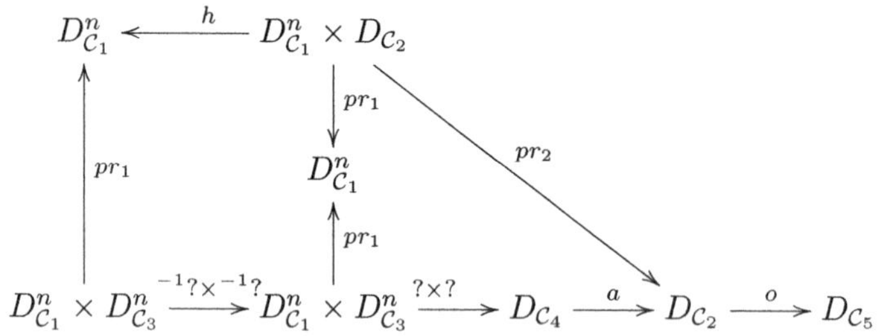

In this setup, the maps are all given by functors between the given categories. An ACN, called *Artificial Conceptual Framework*, is a diagram of ACs.

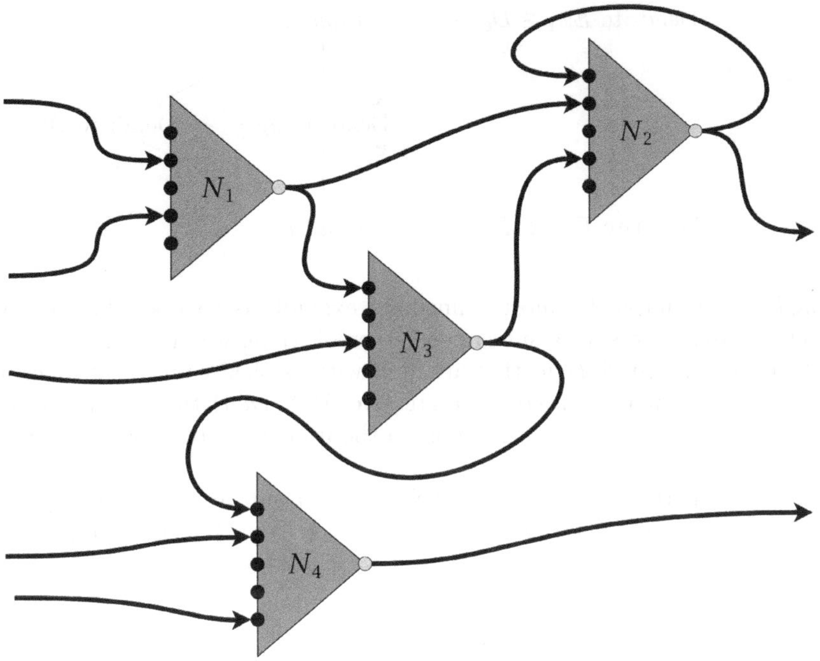

Fig. 4.3: A network of artificial neurons.

The important difference is that in this setup, the input and output of an ACN is a system of 'concepts', i.e., objects in $\mathcal{C}$. We are of course interested in categories involved in the H-jet formalism, but the generic formalism works for general categories $\mathcal{C}$.

The special case of H-jet categories means that an ACN can be used to describe signs with their expressive, significant and content components. More precisely, we shall define a sign $S(ACN)$ that is associated with an ACN to have as its expression the input of ACN, while the signification is specified by the ACN as such, and where the content will be the ACN's output. Let us denote ACNs for $\mathcal{C}$ by ACN($\mathcal{C}$).

Why functors? The classical input and weight are sequences of objects, real numbers in the ANN context. But input and weights may also represent relations among objects. Technically speaking relations are represented by morphisms between objects. This is absent in the ANN approach. Functors solve this deficiency in that they now allow for arguments that are relational.

The semiotic representation of the dendrite of an AC can be done as follows. We have the n inputs Ex_i, which collaborate to generate the dendrite's content:

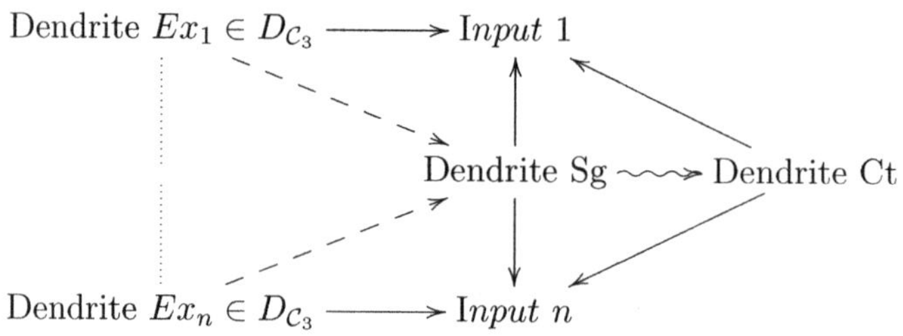

Example 1. The musically most immediate example is to take the category $\mathcal{C} = Den$ of denotators, or a subcategory thereof. Here, we may construct musically meaningful ACN(Den) that are built with musically universal concepts. In particular, we may then conceive creative ACN(Den) that create musical objects, such as standard compositions, in the framework of network architectures!

In this situation, the standard ANNs become a subcategory of ACN(Den): A real number r can be interpreted as a denotator $Den(r)$ of simple type, zero-addressed, with module $\mathbb{R}$, and with empty denotator name, the form being called $ANN2Den$.

4.8 Examples

4.8.1 A Functorial Example with Tensor Products

We take $n = 1$ and $\mathcal{C}_1 = \mathcal{C}_2 = \mathcal{C}_3 = \mathcal{C}_4 = \mathcal{C}_5 = \mathbf{LinMod}_R$, the category of modules and linear maps over a commutative ring R. The map $?\times?$ is the tensor product on every time level: $(M_t, N_t) \mapsto M_t \otimes N_t$, a functorial map. The Hebb map h is the first projection. In particular, if the weight is constant in time with value a ring homomorphism $R \to S$, then $?\times?$ is the classical extension of scalars $M_t \otimes_R S$. The maps a and o are the identities.

4.8.2 A Functorial Example with Simple Denotators

This first example works for a subcategory of the category Den of denotators. We take as $\mathcal{C}_1 = \mathcal{C}_3 = \mathcal{C}_4$ the category of simple denotators in $\mathbb{R}^m$, together with isometric $\mathbb{R}$-linear maps as morphisms. The category $\mathcal{C}_2$ is the category of simple denotators in $\mathbb{R}$, together with linear maps. The functor $?\times?$ is the scalar product $(x_1 \ldots x_n) * (y_1 \ldots y_n) = (x_i^t . y_i)_i$, x^t being the transposed vector. The functor maps an n-tuple (M_i) of isometries on $\mathbb{R}^m$ to the identity, in fact, $(M_1(x_1) \ldots M_n(x_n)) * (M_1(y_1) \ldots M_n(y_n)) = (x_i^t M_i^t . M_i y_i)_i = (x_i^t . y_i)_i$.

The Hebb map h maps $((x_i)_i, y)$ to $(y.x_i)_i$, the morphism $((M_i)_i, \lambda)$ maps to $(\lambda M_i)_i$.

The linear evaluation map a takes $x \in \mathbb{R}^m$ to the scalar product $x^t.q \in \mathbb{R}$ for a fixed non-zero vector $q \in \mathbb{R}^m$. For a linear map $M : \mathbb{R}^m \to \mathbb{R}^m$, we have

$x^t M^t.q = x^t.q.M_2, M_2 \in \mathbb{R}$. We therefore get $M_2 = q'^t.M^t.q$, for a vector q' with $q'^t.q = 1$. Evidently, for $N : \mathbb{R}^m \to \mathbb{R}^m$, we get the corresponding N_2, and $(N.M)_2 = N_2.M_2$.

The evaluation map o can be any scalar multiplication on $\mathbb{R}$.

This situation allows for inputs and weights that are also morphisms between denotators, not only the denotator objects!

4.8.3 Yoneda

A canonical example is the Yoneda Lemma construction: Let $n = 1$, $\mathcal{C}_3$ a category $\mathcal{C}$, $\mathcal{C}_i$, $i = 1, 2, 4, 5$, the set presheaf category $\mathcal{C}^@$ of $\mathcal{C}$. The map $?\times?$ is the map, at time t, $\mathcal{C} \times \mathcal{C}^@ \to \mathcal{C}^@$ sending a pair (X, F) to the filter $@_F X = F \cap @X$, where we denote $@X = \mathcal{C}(?, X)$. The Hepp map is the first projection. Typically, for semiotic contexts, we take the category $\mathcal{C} = \mathcal{LDC}/R[\Sigma^{-1}]$ of H-jets.

4.8.4 The Recursion Theorem

It is evident that ACNs also include the definition by recursion, which is guaranteed by the recursion theorem. Classically set, this theorem deals with maps $r : X^{\mathbb{N}} \to X^{\mathbb{N}}$ such that for two elements $(x_i)_i, (y_i)_i \in X^{\mathbb{N}}$, if $x_i = y_i$ for $i < j$, then $(r((x_i)_i))_j = (r((y_i)_i))_j$, in particular, the image at index 0 is a constant since there are no parts for $i < 0$. The recursion theorem then states that r has a unique fixed point, which is the recursively defined concept. The function r is equivalently defined by a sequence $r_i : X^i \to X$ of well-defined maps, which yield $r_j(x_0, \dots x_{j-1}) = r((x_i)_i)_j$. The theorem's proof is an easy application of the fixed point theorem for topological contractions, see [30, Exercise 164].

With this in mind, we construct an AN $\blacktriangleright^w_N$ with $dim(N) = 1$, $a = o = Id$, and the set category $\mathcal{C}_2 = \mathcal{C}_3 = \mathcal{C}_4 = \mathcal{C}_5 = X$. The category $\mathcal{C}_1$ for weights is $\mathcal{C}_1 = X^{X^{\mathbb{N}}}$, the set of maps $f : X^{\mathbb{N}} \to X$. We fix a weight w as follows: $w_i = \text{const.} = y_0$ for $i \leq 0$, and $w_i = f_i \circ pr_i$, with $pr_i X^{\mathbb{N}} \to X^{i-1}$ the canonical projection, and $f_i : X^{i-1} \to X$ the above map r_{i-1}. The Hebb map h is just the first projection. The map $?\times?$ is the map r as above, defined by the weight map sequence.

This AN yields a unique fixed point, i.e., the input equals the output, meaning that we have a circular ACN $\blacktriangleright^w_N\circlearrowleft$ with a unique solution of its state space, this is the recursion theorem's restatement in terms of ACNs.

4.8.5 RUBATO® Networks

The RUBATO® software deals with networks of rubettes, i.e., components that produce output denotators from a sequence of input denotators, and a sequence of rubette system properties. A rubette network combines a number of rubettes in the evident way, where certain rubette outputs are taken as inputs for other rubettes.

It is an interesting exercise to reinterpret rubettes as ANs, where the dendritic input is the sequence of input denotators, while the weight is the sequence of rubette properties. The AN function $?\times?$ is the rubette's functionality in terms of a Java application. The rubette network then is realized as a ACN on adequate subcategories of *Den*.

Example 2. A typical example from rubette functions is the Riemann matrix of a chord, and, moreover, the path of harmonic functions associated with a sequence of chords. In this case, the category $\mathcal{C}_3$ is the discrete category of objects, which are the sequences of chords. The weight category is the discrete category of sequences of Riemann weights, elements of $\mathbb{R}^{3\times 6}$. The map $?\times?$ associates with a sequence ch_i of chords the sequence of the chords' Riemann matrices. The category $\mathcal{C}_1$ is the discrete category of parameters that define the functions defining the Riemann matrices. The map a associates with the sequence of Riemann matrices the best path, i.e., a sequence of pairs of (a, b) in the Riemann matrix, which are defining a best path. The weight is usually taken to be constant, i.e., the q map is the first projection. The chord sequence as a function of time is usually also taken to be time-independent.

This construction can be generalized to any rubette of a rubato network, generating output denotators from input denotators. And combining such ACs for rubettes, one may construct ACNs for any rubette network.

It remains to be seen what is the advantage of the time variable $\mathbb{Z}$ in this conceptual networks, going beyond the classical ANN's time role.

4.8.6 A Melody Creator

This ACN is defining a melodic denotator of arbitrary length. It is not a standard rubette network, but a recursive construction of infinite length, as parametrized by the time parameter in ANs.

The basic knowledge used here is the complete classification of three-element motives in onset-pitch class space $\mathbb{Z}^{12} \times \mathbb{Z}^{12}$ modulo the full (affine) automorphism group of translations and invertible group elements [27, Ch. 11.13.8]. There are 26 isomorphism classes, they are listed in [27, App. M.3].

This model generalizes the construction of the main melody in Mazzola's *Synthesis* composition [25] as described in [27]. That main melody is a union of all 26 three element motives, see Figure 4.4.

Our construction method runs as follows: The input category is the discrete category $\mathcal{C}_1 = SingOP$ of (finite) lists of single note denotators with onset and pitch coordinates. Such a list stands for a melody. The input is a $\mathbb{Z}$-indexed sequence of such melodies. We moreover ask the input to be eventually and initially constant. The task here is to focus on the construction of the non-constant middle part. The weight is also a sequence of lists of natural numbers whose entries refer to the 26 classes. We suppose that for melody input m_i, the weigh sequence w_i is the sequence of the classes of the melody's three-element parts of successive notes. So, if $m_i = (m_i(1), m_i(2), \ldots m_i(k))$,

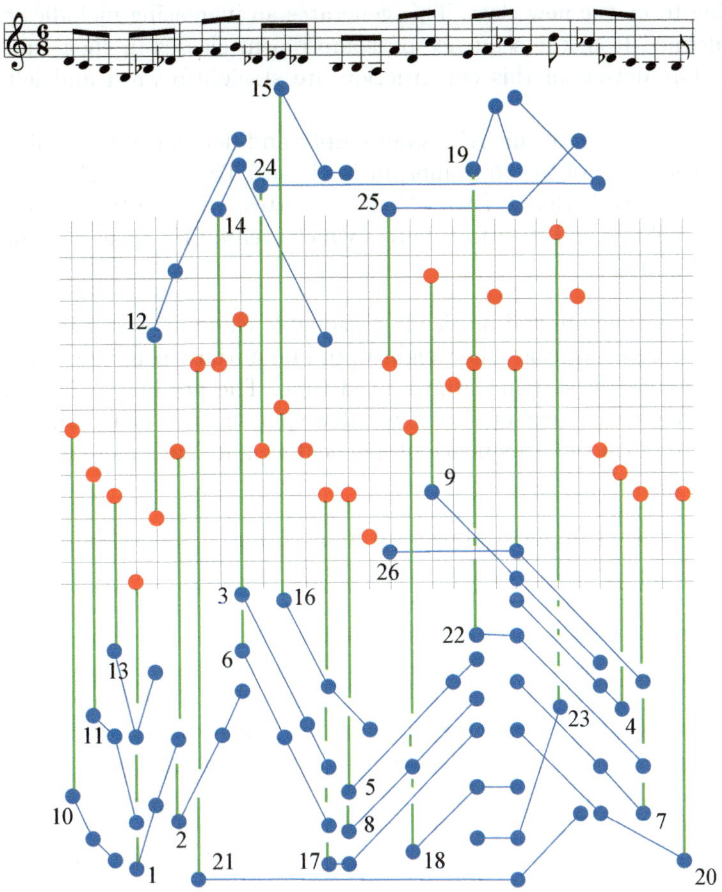

Fig. 4.4: The main melody of the composition *Synthesis* is the union of representatives of all 26 isomorphism classes of three element motives.

then $w_i(1)$ is the class of the three element motif $(m_i(1), m_i(2), m_i(3))$, $w_i(2)$ is the class of the three element motif $(m_i(2), m_i(3), m_i(4))$, etc. The critical map ?×? works as follows.

We first prepend all motif class numbers in a fixed order (independent of the present procedure) of those classes that are not present in the given weight w_i. Then we define the next note that must be appended to the given sequence m_i as follows. We take the last two-note interval of m_i, $(m_i(k-1), m_i(k)$. We then take the first three-element motif class, where there is a representative (out of the generically $12 \times 4 = 48$ candidates) that continues $(m_i(k-1), m_i(k)$ to yield $(m_i(k-1), m_i(k), m(k+1)$, with onset of $m(k+1)$ larger than the last onset of m_i. This is always possible since the retrograde inversion of the last three-element motif of m_i can be appended. We then extend the weight by appending to w_i the new class. This generates an increasing melodic structure that is defined from the sequence of isomorphism classes in their function as weights. The details of this construction are straightforward and left to the reader.

The state space of this AN is not empty and depends on the initial input and the fixed order of the 26 isomorphism classes. To use this AN for an ACN, we may now attach this AN to ANs for rubettes, which alterate the melody to a tonal melody, and then to rubettes which create first species counterpoint from tonal melodies.

Example 3. A concrete and musically reasonable example would be a constant input of a score denotator. The weight would be initially constant $= 1$, q would add 1 to the weight starting from a fixed time. The product is $D^n \times D \to D$ would be the assignment of the metrical weight via the metro rubette. This generates an output of a sequence of metrical weights with increasing minimal local meter lengths.

Example 4. This example, dealing with film scoring, is rather problematic, not because it is not precise, but because its logic should be much more sophisticated than the present state. We are dealing with film score music. The classical approach in film industry is driven by the introduction of music in a "postproduction" phase. This means that the visuals of the film are completely recored, and that from this 'template', the music is generated. Such a musical completion is in itself quite complex. Essentially, the music supervisor first collects 'raw material' according to the visual data. He then transfers his materials to the composer, who, in a final stage records his compositions with adequate orchestras, of course in a tight collaboration with the music supervisor and his staff. This is all done without AI technology—except for the usual sequencer and other digital audio workstation software.

In the following discourse, we don't want to replace the composer's role, but help the music supervisor select basic music material using ACN ideas. Essentially, we shall have a formal representation of the movie's visual part in the role of an ACN weight, together with the input that contains a generic repertory of musical structures. The AN's output will be a selection of musical structures that could yield candidates for the music supervisor's repertory.

We start with the construction of an AN that outputs a selection of musical structures.

- **Weights** The AN's weight is built as follows. We first decompose the visual movie into a $\mathbb{Z}$-parametrized sequence $(A_t)_t$ of *cues*, i.e., of elementary events, which typically describe actions that follow each other along the movie's time line. The first task here is to propose a formal representation of the cues.
  Our approach starts with a simple cue architecture:
  Cue = Agent+Action+Objective.
  The Agent can be

Human/Animal/Plant/Nonliving.
Each of them can be one of the pronoun specifics:
I/You/He(She,It)/We, You, They.
The Action can be
Aggressive/Peaceful/Connecting/Disconnecting/Indeterminate/Surprising/... Expected.
The Objective can be, again,
Human/Animal/Plant/Nonliving.
For example, the cue for "John kills Mary." would be represented by
Human:He/aggressive/Human:She.
This information should be complemented by an onset and duration information in *hours/minutes/seconds.*
Formally, we may shape these attributes by forms and their denotators. We define the form *Cue* as a limit of *Agent*, *Action*, *Objective*, and *Duration*, see Figure 4.5. The factor *Duration* is a limit of simple forms *Onset* and *Offset*, both with $\mathbb{R}^3$ as coordinator for hours, minutes, and seconds. *Agent* would be a colimit of limit forms *Human*, *Animal*, *Plant*, *Nonliving*, which have simple factors *value* with $\mathbb{R}$ as coordinators and the limit form *Person* that has the $\mathbb{R}$-coordinate forms *I*, *You*,...*They* as factors. Similarly, we define the form *Action*, which is a colimit of simple forms *Aggressive*, *Peaceful*,... *Expected* with $\mathbb{R}$ as coordinator. The factor *Objective* has the same structure as the one for *Agent*. This setup enables the assignment of real number 'weights' to each of the simple forms, respectively. Therefore a *Cue* denotator defines 'weights' for each of its components. It would also be reasonable to limit these weights to the closed real interval $[0, 1]$.

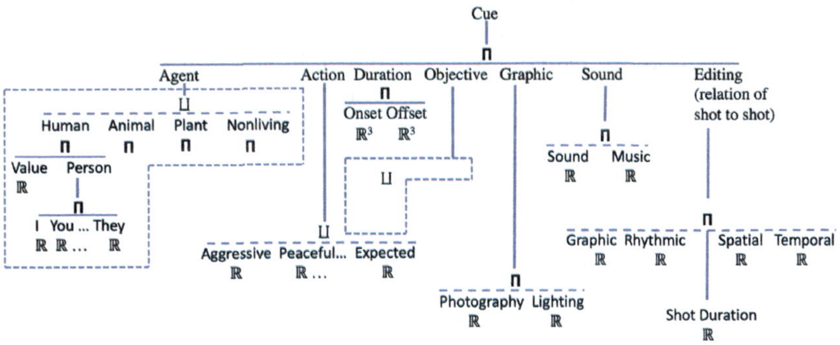

Fig. 4.5: The Form Cue.

- **Musical Structures** These weights, as represented by *Cue* denotators, are supposed to act upon musical structures and thereby selecting collections of such structures, which are susceptible for candidates to the composer. To

be clear, these selections are not compositions, but musical raw materials to be used by the composer(s) for the construction of a concrete musical text.
The musical structures from which the selections will be taken are the input of our AN. The output will be a subset of the input, which is defined from the interaction with the above *Cue* weights.
We exhibit the following types of musical structures:
Melodies, Harmonies, Rhythms, Tempi, Orchestrations.
The input will be, for each of these types, a set of examples, e.g., a set of melodies. Each type's set will be a powerset denotator containing the set of structures. Let us call these powerset forms *MelSet*, *HarmSet*, *RhythmSet*, *TempSet*, and *OrchSet*, referring to the coordinator forms *Mel*, *Harm*, *Rhythm*, *Temp*, and *Orch*, respectively.
The input would then be a sequence/list of such powerset denotators, one for each type. The output would be a corresponding list of subsets of these powerset denotators, the selection as generated from the list's interaction with the *Cue* weight.
This selection process works as follows: For every melody, harmony, rhythm, tempo, or orchestra in one of the list's sets, the given *Cue* acts upon this instance and then decides upon its in- or exclusion in the selection subset. This logic implies that our musical structures have to reveal a weight with respect to the dramatic *Cue* denotator, i.e., *a dramatic interpretation of a musical structure is needed.* This evidently is the core problem in this AN mechanism.
This problem is evidently *not* the general situation in music semiotics. The dramatic interpretation of music is a very special perspective, which cannot replace musical analysis that doesn't refer to dramatic categories. This fact is essential because the 'neutral' display of musical structures is not sufficient in the film score context.
This means that we cannot expect a direct interaction of structure and *Cue*. The structure needs a dramatic semiotic enrichment. Or else should we propose a method that 'dramatizes' musical structure with a view to the *Cue*. At present, this second option reaches beyond our competence. We therefore make the hypothesis that a dramatic specification of any structure in our sets of the input list is given.
Such a dramatically interpreting added value to musical structures should enable a link to both the general musical structure and the above *Cue* form of dramatic components. These two agents would interact to yield a numerical/logical result which in- or excluded the given structure in or from the selection subset.
The evident solution is to attach to every musical structure S a $Cue(S)$ denotator specific to that structure. Then one may attribute to the pair $(Cue(S), Cue)$ a real number value $Cue(S) *_{Type} Cue$, which is defined by a bilinear form $*_{Type}$ specific to the type, i.e., type = melody, harmony, rhythm, tempo, orchestration. For type "melody" one would give a high

weight to the "Agent" part, and a low weight to "Action", "Objective", etc. The value $Cue(S) *_{Type} Cue$ would define whether the structure S is included in the selection subset or not.
This means that we set up limit forms $Struture + StructuralCue$ with factors $structure\ S$ and $StructuralCue(S)$. For the colimit components, one would set $Cue(S) *_{Type} Cue$ to be zero if different colimit cofactors were at stake.

- **Inter-Cue Logic** The above evaluation procedure produces output structures that are independent of each other on every time $t \in \mathbb{Z}$. But the $Cue(t)$ sequence over time t is not random, there always exists a logical connection inherited from the film's narrative. The question from this 'logical spine' is, how to connect the outputs in time as a consequence of this 'spine'. This means to define such a *logical spine* as a structure deduced from the $Cue(t)$ sequence, which may help select musical structures in a connected style from the above selected subsets. Evidently, this spine selection would be accomplished by another AN that takes the present output as an input, but also copying the Cue weights to new weights.
The question of how to define/exhibit a logical spine from the $Cue(t)$ sequence is open at present. Such a logic should take care of the continuities and disruptions of consecutive $Cue(t)$.

5

Semantic and Expressive Topology

Summary. This chapter presents a (classical) topology on H-jet collections that relates to semantic aspects. Dually, we shall introduce an expressive topology.

– Σ –

5.1 Semantic Topology of H-jets

The semiotic analysis deals with the global configuration of H-jet relationships. The semantic potential of a given H-jet Y can be described by the full subcategory $Y@/$, the objects and morphisms that can be reached from Y. We therefore define the *semantic topology* $\mathcal{C}^{\mathrm{sem}}$ on a (small) category $\mathcal{C}$ as follows. The open sets are sets U of morphisms in $\mathcal{C}$ which are content-stable. This means that if f is a morphism in U, then the category $dom(f)@/$ defined by the domain $dom(f)$ of f is a subset of U.

Evidently, $\mathcal{C}$ is open in $\mathcal{C}^{\mathrm{sem}}$, and so is the empty set. Moreover, any union of open sets and every intersection of open sets are open. The latter is true since for $Y@/ \subset U_i$, all U_i of a family (U_i), we have $Y@/ \subset \bigcap_i U_i$. Every morphism f has a minimal *elementary (semantic)* open neighborhood, namely $dom(f)@/$. The latter is open because for $g \in dom(f)@/$, $dom(g)@/ \subset dom(f)@/$. But it is not true in general that the intersection of elementary open neighborhoods is again elementary.

An open set U in $\mathcal{C}^{\mathrm{sem}}$ is also a full subcategory. In fact, if a morphism f is in U, then so is $dom(f)@/$, and therefore all morphisms with domain $dom(f)$. If $f : Y \to Z, g : Z \to W$ are two morphisms in U, then Y is in U, and therefore $Y@/$, and therefore $g \circ f$.

The *open points* are those, whose elementary neighborhood is just these points, i.e., the final objects. The *closed points* are those which are in no other's elementary neighborhood, i.e., those which have no "semantic value" for any other points, the "meaningless" points. More generally, the closure $\overline{\{X\}}$ of a

G. Mazzola et al., *Functorial Semiotics for Creativity in Music and Mathematics*, Computational Music Science, https://doi.org/10.1007/978-3-030-85190-3_5

point X is the set of points for which X is in their elementary semantic neighborhood.

Finally, if $F : \mathcal{C} \to \mathcal{D}$ is a functor, then it is always a continuous map $F^{\text{sem}} : \mathcal{C}^{\text{sem}} \to \mathcal{D}^{\text{sem}}$, i.e., *We have a functor*

$$\text{sem} : CAT \to \mathbf{Top}.$$

One should now reconsider statements of categorical nature in terms of this semantic topology. For example: How far is the functor sem away from reflecting homeomorphisms? In particular, what is the difference between the homeomorphism class of $SEMCLASS(X)$ and the category-theoretical $SEMCLASS(X)$? What are the fundamental groups in $\mathcal{C}^{\text{sem}}$? More generally: What is the structure of gesture categories for the semantic topologies? We shall make some very first statements below about Čech cohomology in these topologies.

Example 5. If $\mathcal{C}$ is the path category of a digraph, then, for an object (point) p of $\mathcal{C}$, the elementary neighborhood of p is the subcatgeory of all points q that can be reached from p.

Example 6. If $\mathcal{C}$ is the category of commutative rings, then the closure of the zero ring is the entire category. More generally, for a ring R, its closure is the set of all rings S, for which R is a S-algebra.

Example 7. If $\mathcal{C}$ is an abelian category, then the semantic topology is the indiscrete topology.

Remark 4. The traditional Yoneda context deals with functors F in $\mathcal{C}^{opp,@} = \mathcal{C}^{opp}@CAT$ which can be investigated about their representability, i.e., $F \overset{\sim}{\to} @Y$. In our semantic topology context, the analogous situation is as follows: We are looking at the category $S(\mathcal{C})^{opp}@CAT$ of functors on the category $S(\mathcal{C})$ of open subsets plus inclusions of $\mathcal{C}$ with values in CAT. Such a functor F will be called *semantically representable* if there is a category X in CAT, such that $F(U) = U@X$, the category of functors $f : U \to X$ with their natural transformations as morphisms. *The interesting thing about this construction is that now, we recover the original presheaf structure for general categories with their semantic topology.*

Example 8. If we consider the subcategory $Iso(\mathcal{C})$ of isomorphisms on the $\mathcal{C}$, then the functor $F : S(\mathcal{C})^{opp} \to CAT$ evaluating to $Iso(U)@G$ for a group (viewed as a one-point category) is not semantically representable in general.

Example 9. Looking for example at Beethoven's piano sonatas, numbered by #1 to #32, we may look at the H-jets J_n=Ex(Sonata # n) $\to$ Sonata # n, and we may consider the open covering $U(BeethovenSonatas)$ by the categories of J_n. This defines a cohomology $H^*(R, U(BeethovenSonatas))$. This formulas are expressing the sonata categories and their intersections, together with their cohomological modules, what we call *Beethoven Sonata cohomology.*

5.1.1 Limits

The elementary neighborhood $Y@/$ of a H-jet Y describes the full semantics of Y. Often, this is more than necessary to characterize Y up to isomorphisms. It may in fact happen that $Y \stackrel{\sim}{\to} \lim \mathcal{D}$ is the limit of a diagram $\mathcal{D} = (Y_\iota)\ldots$ of H-jets. This diagram is part of $Y@/Y$, the co-slice category of Y that gives rise to $Y@/$. In other words, the elementary neighborhood $Y@/$ comprises the diagram $\mathcal{D}$ whose limit is Y.

If Y is such a limit, this characterizes Y via its semantic structure as given by $\mathcal{D}$. This is not always the case: a H-jet Y may be "more" than its semantic potential as given by $Y@/$, its identity cannot be deduced from its contents; e.g. the H-jet "God".

Many H-jets however do fulfill this limit characteristic, let us call them *limited*.

5.2 Expressive Topology of H-jets

Dually to the semantic topology on category $\mathcal{C}$, one may consider the semantic topology on $\mathcal{C}^{opp}$, which we for evident reasons call the *expressive topology* on $\mathcal{C}$, and we denote it by $\mathcal{C}^{\text{ex}} := (\mathcal{C}^{opp})^{\text{sem}}$. Evidently, the elementary neighborhood of H-jet Y in the expressive topology is defined by the set of objects of the closure of Y in the semantic topology.

Part III

Semantic Math

6

Concept Mathematics

Summary. Part III deals with the very definition of conceptual mathematics and then a discussion of first attempts to generate a systematic approach to conceptual mathematics.

– Σ –

This Part III is not restricted to the usual definition and theorems, but extends to an investigation of a new type of mathematics, which we call "concept mathematics". In Section 10.4.1, we shall discuss more systematically the need for such a type of mathematics. Here we just want to prepare the reader for this transition.

The origin of our proposal stems from the above introduction and investigation of categories of semiotic signs, represented by H-jets in our terminology. This setup enforces a search for the semiotic anatomy of mathematical conceptualization. Such a necessity follows from the simple, but painful insight that mathematicians (and everybody, but in a less pronounced style) don't know what they are doing, the Goldbach conjecture being the example *par excellence*: simple concepts enable unsolvable questions! Conceptual mathematics should focus on this mysterious movement of the mathematical body when it adds concepts. Calculations are not the core of mathematics, they are rather an excuse and trick to circumvent the crucial questions underlying the very problem of problem solving.

Part III will be devoted to potential candidates of the body's organs of conceptual mathematics. This body is above all a growing entity, adding new mathematics is the crucial operation, where our lack of control becomes visible. In other words, we shall discuss the process of creativity underlying the strategies of solving mathematical problems.

G. Mazzola et al., *Functorial Semiotics for Creativity in Music and Mathematics*,
Computational Music Science, https://doi.org/10.1007/978-3-030-85190-3_6

7

Yoneda

Summary. We discuss some global consequences of Yoneda's Lemma and its philosophy.

– Σ –

7.1 Yoneda's Lemma as a Semiotic Statement

Nobuo Yoneda was a Japanese computer scientist and prouncounced his now celebrated "Yoneda Lemma" to Saunders Mac Lane at a Café at Gare du Nord in Paris 1954 before his departure back to Japan. The lemma is not difficult to prove, but it is a deep conceptual result about categories.

Lemma 4. *(Yoneda Lemma). If $\mathcal{C}$ is a category, we consider the category $\mathcal{C}^{@}$ of presheaves over $\mathcal{C}$, whose objects are the contravariant functors (presheaves) $F : \mathcal{C}^{opp} \to$ **Ens**, together with the natural transformations between presheaves as morphisms. Then we have a functor Yoneda : $\mathcal{C} \to \mathcal{C}^{@}$, which associates with every object Y in $\mathcal{C}$ the presheaf $@Y : X \mapsto X@Y$. This functor is fully faithful, more precisely, we have a functorial isomorphism*

$$@Y@F \stackrel{\sim}{\to} Y@F$$

($Y@F$ being the evaluation of F at Y) for every object Y of $\mathcal{C}$ and every presheaf F, i.e., the natural transformations from $@Y$ to F are one-to-one with the values $Y@F$ of F at argument Y. In particular, if $F = @X$, then $Y@X \stackrel{\sim}{\to} (@Y)@(@X)$.

The semiotic interpretation of this lemma is that to know the isomorphism class of an object Y of $\mathcal{C}$ it is necessary and sufficient to know its functor $@Y$. This means that Y is identified by the 'integral' of all 'perspectives' $f : X \to Y$. The mathematical content of Y is the 'integration' of all 'views' from the arguments X in the given category. This is what we called the "Yoneda philosophy" in [24].

G. Mazzola et al., *Functorial Semiotics for Creativity in Music and Mathematics*, Computational Music Science, https://doi.org/10.1007/978-3-030-85190-3_7

This interpretation is the basis of our approach to categories when viewed as topological spaces with their "semantic topology". We don't agree with Pierre Cartier[1] who called the lemma "le bon usage d'une tautologie", the proof being simple—but the conceptual statement is not.

For our needs, we can state a more general Yoneda Lemma. To this end, we embed the category **Ens** in the category CAT, $SC : \mathbf{Ens} \to CAT$ by interpreting a set as a discrete category (only identities as morphisms).

Lemma 5. *(CAT Yoneda Lemma). If $\mathcal{C}$ is a category, we consider the category $\mathcal{C}^{CAT}$ of CAT(-valued) presheaves over $\mathcal{C}$, whose objects are the contravariant functors (CAT presheaves) $F : \mathcal{C}^{opp} \to CAT$, together with the natural transformations between CAT presheaves as morphisms. Then we have a functor $Yoneda : \mathcal{C} \to \mathcal{C}^{CAT}$, which associates with every object Y in $\mathcal{C}$ the presheaf $CAT(Y) : X \mapsto X@Y \mapsto SC(X@Y)$. This functor is fully faithful, more precisely, we have a functorial isomorphism*

$$CAT(Y)@F \overset{\sim}{\to} Y@F$$

($Y@F$ being the evaluation of F at Y) for every object Y of $\mathcal{C}$ and every CAT presheaf F, i.e., the natural transformations from $CAT(Y)$ to F are one-to-one with the values $Y@F$ of F at argument Y. In particular, if $F = CAT(X)$, then $Y@X \overset{\sim}{\to} CAT(Y)@(CAT(X)$.

The proof is the same as the proof of the classical Yoneda Lemma [14, p. 20], just replace **Ens** by CAT.

7.2 The Bidual Lifting of the Yoneda Construction

The relationship between the semiotic category $\mathcal{LDC}_R^\Sigma$ and the category CAT of categories is a special case of a much more general fact, which sheds light upon the Yoneda construction. To this end, sets can be view as very special categories: A set S is viewed as a discrete category, the objects of the category S are its elements, and the identities of these objects are the only morphisms of S. Functors between these sets qua categories are the traditional functions between sets. This implies that we may view the Yoneda embedding $\mathcal{C} \rightarrowtail \mathcal{C}^@$ having the codomain $\mathcal{C}^@$ as a category of functors $F : \mathcal{C}^{opp} \to \mathbf{Ens} \to CAT$, where sets are interpreted as discrete categories. This is our new interpretation of the Yoneda construction, it is an embedding $\mathcal{C} \rightarrowtail \mathcal{C}^{opp*}$, where $\mathcal{C}^* = \mathcal{C}@CAT$ is the "dual" category of $\mathcal{C}$.

This being, we consider the bidual category $\mathcal{C}^{**}$. We have the evident bidual functor $Bidual : \mathcal{C} \to \mathcal{C}^{**}$, where $Bidual(X)(F) = F(X)$ for $F \in \mathcal{C}^*$.

We first have a general theorem:

[1] Although we agree with most of what he says.

Theorem 1. *The functor Bidual is an embedding of any category $\mathcal{C}$ as a non-full subcategory of functors with values in CAT.*

Its proof will be given below.

The important question is first about the difference between the Yoneda embedding and the bidual functor. Let us compare this one with the Yoneda embedding:

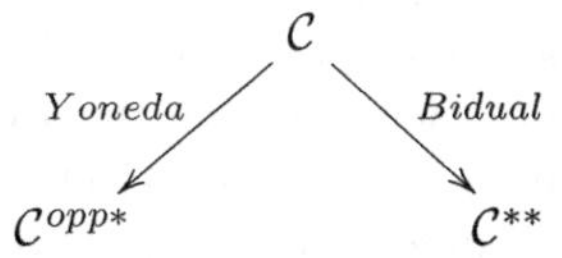

The two maps, $Yoneda$ and $Bidual$, have an astonishingly simple relation. In fact, consider the covariant Yoneda embedding $Yoneda^{opp} : \mathcal{C}^{opp} \to \mathcal{C}^* : X \mapsto X@$. Then we have its dual functor $YB : \mathcal{C}^{**} \to \mathcal{C}^{opp*}$. With this, we have

Theorem 2. *With the above notations, there is a commutative diagram*

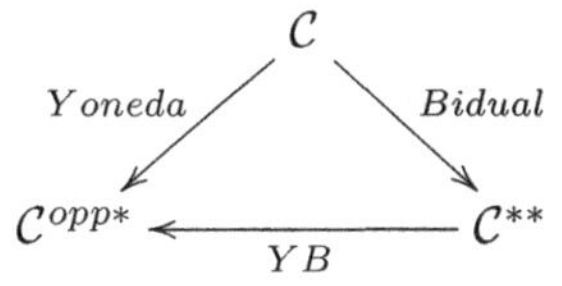

Proof For an object X in $\mathcal{C}$, we have

$$YB(Bidual(X))(Y) = Bidual(X)(Y@) = Y@X = Yoneda(X)(Y).$$

For morphisms the same fact holds.

In particular, since $Yoneda$ is injective, so is $Bidual$, this diagram proves the injectivity of the $Bidual$ as claimed in the above theorem. One could state that this factorization of the Yoneda functor brings together covariant and contravariant aspects of morphism sets $X@Y$.

Let us prove next that YB is surjective. Here are the arguments: Taking a functor $F : \mathcal{C}^{opp} \to CAT$, the question would be if there is a functor $G : \mathcal{C}^* \to CAT$, such that $F = G \circ Yoneda^{opp}$. If $F = @Z$, then the evaluation at Z, $?(Z) : \mathcal{C}^* \to CAT$ solves the problem: $@Z = ?(Z) \circ Yoneda^{opp}$. Now, it is well known that every functor $F : \mathcal{C}^{opp} \to CAT$ is the colimit of a diagram of representable functors: $F = \mathrm{colim}_i @Z_i$. The canonical candidate therefore would be $G = \mathrm{colim}_i ?(Z_i)$. We have to check $\mathrm{colim}_i @Z_i = \mathrm{colim}_i ?(Z_i) \circ Yoneda^{opp}$. There is a canonical natural transformation

$$\mathrm{colim}_i @Z_i \to \mathrm{colim}_i ?(Z_i) \circ Yoneda^{opp},$$

but it is not clear, whether this is an isomorphism. If we go into the natural transformations involved in such a diagram of functors, we see colimits of this shape for every argument Y of $\mathcal{C}$:

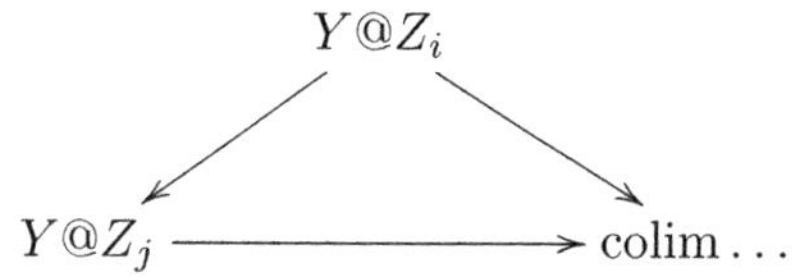

This is a colimit diagram of categories. In the left part of the previous arrow, this is the evaluation of $@Z_i$ at Y, in the right part, it is the evaluation of $?(Z_i)$ at $@Y$. But both interpretations yield the same colimit triangle. Therefore these two colimits are the same, and the above arrow is in fact an isomorphism. This implies that the YB map is a surjection, i.e., every Yoneda functor in $\mathcal{C}^{opp*}$ stems from a bidual functor in $\mathcal{C}^{**}$, i.e., we have this commutative diagram:

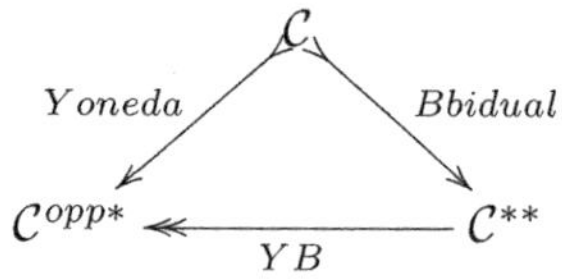

The embedding $Bidual : \mathcal{C} \rightarrowtail \mathcal{C}^{**}$ enriches every object X (well: H-jet in our context) by its bidual representation.

Remark 5. There is a deep hidden rationale of the Yoneda embedding that relates the structural mathematical approach of categories to the object-oriented set-theoretical approach. In fact, the calculus of categories, which abstracts from objects, is not in itself capable of doing mathematics alone. The Yoneda embedding overcomes this deficiency by replacing abstract categories via the presheaves associated with objects, together with the natural transformations replacing abstract morphisms. In fact, instead of dealing with abstract categories, Yoneda enables a mathematics of objects when dealing with categorical entities. The presheaf is $@X$ is no more that abstract original entity X, it is now a system of sets, i.e., of objects that are connected by set-theoretical functions. Yoneda reconstructs the obect-oriented environment to control abstract categorical entities.

When we envisage a transition from structural mathematics to conceptual mathematics, we need a tool that plays the role of Yoneda for the transition to concepts. Let us call this hypothetical tool the *Conceptual Yoneda Lemma*. It would enable a calculus with categories that controls the conceptual framework by an explicit tool on the level of categories. It is probable but not clear at present that the bidual functor could play this role. The evident difference to Yoneda is that the bidual is not fully faithful anymore.

7.3 A Concrete Separating Functor of Semantic Significance

The previous discourse is a very general one and needs to be understood in more concrete terms, namely focusing on concrete functors $\mathcal{C} \to CAT$ for our semiotic category $\mathcal{C} = \mathcal{LDC}^{\Sigma}_{R}$. The overall perspective of the *Bidual* embedding is that it gives the semiotic category a representation in terms of "coordinate functions", those functors $F : \mathcal{C}^* \to CAT$, which may separate different objects (H-jets in our case).

Here is a first candidate for such a separting "coordinate function":

Proposition 3. *For an object X of $\mathcal{C}$, let $X@/X$ be the full co-slice subcategory generated by all morphisms $X \to Y$. Let HY be the functor $\mathcal{C} \to CAT : X \mapsto HY(X) = (X@/X)^{@}$ that maps X to the category of CAT-valued presheaves on $X@/X$. If $f : X \to Y$ is a morphism in $\mathcal{C}$, then $HY(f)$ is induced by the canonical co-slice functor $? \circ f : Y@/Y \to X@/X$.*

The evaluation $X \mapsto Bidual(X)(HY)$ is an injective functor $\mathcal{C} \to CAT$. It separates objects and morphism sets. The symbol HY stands for "Hjelmslev-Yoneda".

Proof This functor has our critical property since, if $HY(Y) = HY(Z)$, then also $Y@/Y = Z@/Z$. But then all objects, the co-sliced $f : Y \to W$ and $g : Z \to V$ must coincide, in particular their domains Y and Z must coincide. The same argument holds for the injectivity on morphisms.

The separating single Hjelmslev-Yoneda functor HY is a "coordinate function" with values in CAT that creates a faithful projection of $\mathcal{C}$ into the category CAT of mathematical structures. In the case $\mathcal{C} = \mathcal{LDC}^{\Sigma}_{R}$ which is our focus, the evaluation is a semantic one: the category $HY(X)$ comprises all semantic targets of H-jet X, it is a *semantic evaluation of* X. This coordinate function associates a categorical coordinate with a semantic entity. This proves that *there are enough categorial values to parametrize faithfully semantic entities.*

7.3.1 Semantic Classes

We should be aware that the injection *Bidual* is far from being surjective. More concretely, the functor HY maps $Y@Z$ into $HY(Y)@HY(Z)$, but there are functors $HY(Y) \to HY(Z)$, which do not stem from $Y@Z$. For example, the categories $Y@/Y$ and $Z@/Z$ might be isomorphic without any existing morphism $Y \to Z$, yielding an isomorphism $HY(Y) \stackrel{\sim}{\to} HY(Z)$. Semiotically speaking, the global semiotics of Y might be isomorphic to the global semiotics of Z without any H-jet morphism between Y and Z.

We now want to describe $HY(Y)@HY(Z)$. First, taking the category of all small categories $X@/X$ and their connecting functors, the construction $HY(X) = (X@/X)^{@}$ is the Yoneda functor, which is fully faithful, i.e., we have to investigate the structure of $(Y@/Y)@(Z@/Z)$.

Let $F : Y@/Y \to Z@/Z$ be a functor. It maps the initial object $Id_Y : Y \to Y$ to an object $F(Id_Y) : Z \to F_1$. If $u : Y \to W$ is a general object in $Y@/Y$, then the morphism $u : Id_Y \to u$ means that $u = u \circ Id_Y$. Its image under F is the factorization $F(u) \circ F(Id_Y) = F(u)$, where the left $F(u)$ is the image of the morphism, while the right hand $F(u)$ is the image of the object $u : Y \to W$.

Therefore, the image of an object u is the composition $F(u) \circ F(Id_Y)$, where $F(u)$ here is the image of the morphism $u : Y \to W$. Therefore the image of objects in $Y@/Y$ are defined from: first the image $F(Id_Y)$, and second the image of morphisms on the full category $Y@/$ (without co-fibrations) under the functor[2] $F^* : Y@/ \to Z@/$ underlying F. This means that

Proposition 4. *Any functor $F : Y@/Y \to Z@/Z$ is defined by any functor $F^* : Y@/ \to Z@/$ plus any morphism $F(Id_Y) : Z \to F_1$ in $Z@/$, where F_1 is in the image of F^*.*

This means that our functors F are functors F^*, and with each such F^*, the set of morphisms $Z \to Im(F^*)$. Observe that the functors we had constructed from morphisms $f : Z \to Y$ are coming from first the inclusion morphism $F^* : Y@/ \subset Z@/$ and the morphism $F(Id_Y) = f$. But again, the general functors F^* are not related to any morphisms connecting Y to Z. The only morphism needed here is from Z to any object in $Im(F^*) \subset Z@/$.

In other words, we have the surjection (with a cartesian square)

$$\begin{array}{ccc} (Y@/Y)@(Z@/Z) & \longleftarrow & Z@Im(F^*) \\ \downarrow{\scriptstyle *} & & \downarrow \\ (Y@/)@(Z@/) & \longleftarrow & F^* \end{array}$$

whose fiber over F^* is $Z@Im(F^*)$. The fact that it is a surjection follows from the very definition of $Z@/$, every element of $Z@/$ is the codomain of a morphism starting at Z.

It is remarkable that the coordinate functor HY, which originally connects morphisms $Y \to Z$, now connects semantic categories $Y@/$ and $Z@/$, which may have no morphism that connects them. This being, we now are able to conceive isomorphism classes of semantic categories $X@/$. This means that we now understand the semantic potential of a H-jet.

This *global semantic classification* of H-jets is the most important next topic of functorial semiotics. Let us call *semantic class* of a H-jet X, or more generally an object X of category $\mathcal{C}$, the isomorphism class of $X@/$, and denote it by $SEMCLASS(X)$. This object is the categorical classifying image of the separating functor HY at X.

Let us give a simplified small example of such a situation. We focus on the 'pure' sets without any other semiotic context. Sets gain their significance and content by their reference to their elements, the relation that is opposite to the $\in$ relation, and which was introduced by mathematician Paul Finsler [10]. For example the set $1 = \{0\}$, with $0 = \emptyset$ has one arrow $1 \to 0$ to its

[2] Attention, here $Y@/$ denotes the full category defined by the objects which are codomains of morphisms starting at Y, not as previously similarly denoted the functor category in the Yoneda context, but this should not be confusing.

single element, that's what 1 qua set means. And 0 points nowhere, it is a terminal H-jet. In this understanding, there are two well known definitions of natural numbers. The first is that 3 is the set with a unique arrow $3 \to 2$, while $2 \to 1$ is the unique arrow of 2, set-theoretically this translates to $3 = \{2\}, 2 = \{1\}$ and to the corresponding diagram of H-jets $3 \to 2 \to 1 \to 0$. The second representation of natural numbers is given by the ordinal number approach, where $0' = 0, 1' = 1, 2' = \{0', 1'\}, 3' = \{2', 1', 0'\}$. Here we have the corresponding element-relational arrows $1' \to 0', 2' \to 1', 2' \to 0', 3' \to 2', 3' \to 1', 3' \to 0'$. If we take $X = 3$ and $Y = 3'$, the set-(theoretically restricted) generated categories $X@/$ and $Y@/$ are isomorphic, but no arrow between X and Y is given, i.e., $SEMCLASS(3) = SEMCLASS(3')$.

8

Semantic Representations

Summary. This chapter opens the question about the semantic "loading" of a mathematical concept. This representation relates to the categories of H-jets, where concept are conceived as "sources" of a semantic extension or, dually, as "sinks" of an expressive extension.

– Σ –

The semantic (or dually: expressive) representation of a mathematical concept is evidently the starting point of any valid investigation: to understand the meaning of a concept. But this superficial evidence turns into a mysterious multitude of aspects that the concept's meaning displays, and moreover the multitude of aspects that are hidden from the surface, and which are conceptually arcane, not evident at all, neither as supporting instances of the concept's problematic aspects, nor as core attributes of the concept's anatomy.

This situation is even more dramatic as soon as one wants to manage the totality of the concept's semantic loading: it is no given explicitly, there is no "complete" knowledge of what a concept means, not even if one limits one's focus to "pure" mathematics. The category of H-jets is full of potential arrows, which are added in creative actions. For example, the Fermat conjecture enables the famous interpretation in terms of coefficients of an elliptic curve, a decisive semantic perspective that was not known until the last four decades of the 20th century!

Therefore the topic of a semantic representation of a H-jet relates to an open category, i.e., to a category, whose arrows are not explicitly known at every moment of its consideration. This is a radically different situation as a category is supposed to be a "static" thing, which we can still suppose to be the case, but the collection of arrows needn't be accessible without restriction. This looks like something exotic, but it is quite standard in mathematics. For example, the category of modules comprises all possible modules, which are far from known at any moment. Nevertheless, the arcane position of semantic instances of a H-jet is more dramatic than for classical mathematical categories since the management of concepts effectively needs the complete knowledge, not

G. Mazzola et al., *Functorial Semiotics for Creativity in Music and Mathematics*, Computational Music Science, https://doi.org/10.1007/978-3-030-85190-3_8

only the potential one: we want to solve problems, answer question, that a given concept incites.

The conjectured semantic representation *in praesentia* must be constructed, it must be drawn up from the *orcus* of hidden contents to the managble surface of mathematical calculations.

9

Čech Cohomology

Summary. We discuss two approaches to Čech cohomology: function spaces for global filters, and functorial cohomology associated with the semantic topology.

– Σ –

9.1 Spaces of Functions

To define semiotic function spaces on filter functors, we recall the construction of affine function spaces for global compositions in the musical classification theory, see [27, Chapter 15] for details. For a simplex σ of an A-addressed global composition (A an R-module for a commutative ring R), its function space is the R-module of morphisms $f : \cap\sigma \to A@_R R$ into the R-module $A@_R R$ of affine homomorphisms $a : A \to R$; $\bigcap\sigma$ denotes the intersection of all members of the simplex σ. Such a morphism f is by definition the restriction of an affine homomorphism of the R-module M underlying the local composition $\cap\sigma$.

In our context an analogous construction can be provided as follows. There are no modules in the general framework of functorial filters F/Y, but as every such filter is based upon the base functor $@Y$, it gives rise, similar to a local composition, to a subfunctor $@_F Y \subset @Y$. A local composition would correspond to such a subfunctor, and the carrier module in music theory would correspond to the representable functor $@Y$. Given a commutative ring S, we would then define a S-*coordinate function* c on F/Y as being the restriction $c : @_F Y \subset @Y \overset{C}{\to} S$ of a functor morphism C into the constant functor S. The set $\Gamma(F/Y, S)$ of such coordinate functions is evidently an S-module. The condition of affinity in the musical context is now replaced by the condition that c must be a functor. Examples of such functions are easily found: Take any morphism $m : Y \to S$, S being a sign for the ring S, then we have the associated functor $@m : @Y \to @S$. Composing this with any functor morphism $@S \to S$, for example a constant morphism, yields a function. This defines a

G. Mazzola et al., *Functorial Semiotics for Creativity in Music and Mathematics*,
Computational Music Science, https://doi.org/10.1007/978-3-030-85190-3_9

map $F@S : Y@S \to \Gamma(F/Y, S)$. Moreover, if $G/Y \subset F/Y$, then we have the S-linear restriction function

$$\Gamma(F/Y, S) \to \Gamma(G/Y, S).$$

In the special case of $S = \mathbb{Z}_2$, we would consider functions $f : @_F Y \to \mathbb{Z}_2$ as truth functions for semantic values in Y of the filter F/Y, interpreting $\mathbb{Z}_2$ as the expression of the sign, whose content would be the truth set TF introduced in section 4.5.2.

This setup enables us to look for the function module complexes $M(F_i/Y)$ (also called coefficient systems in algebraic topology) associated with the simplicial complex of the covering defined by a family (F_i/Y) of filters over Y. The value for a simplex σ of the covering defined by the family (F_i/Y) is the module $M(F_i/Y)(\sigma) := \Gamma(\bigcap_{i\in\sigma} F_i/Y, S)$. The module of global sections of such a module complex is defined by

$$\Gamma M = \lim_{\sigma} M(F_i/Y)(\sigma),$$

the limit of the complex over the simplices of the family (F_i/Y). Evidently, the module complex $M(F_i/Y)$ is flasque, i.e., all functions on zero-dimensional simplices stem from global functions in ΓM.

The general theory of coefficient systems enables the standard construction of cochains of functions, see [27, Section 16.1.2] for details. This entails that for any family (F_i/Y) of filters over Y, we have the cohomology S-modules $H^k(S, F_i/Y)$ that are derived from singular simplices with values in the function modules.

9.1.1 Representing Filters within Function Space Functors

In the classification theory of global musical compositions, there is a fundamental construction of such compositions using spaces of affine functions [27, Section 15.2.2]. This can also be achieved—mutatis mutandis—in our semiotic context.

To begin with, observe that for a given filter F/Y, the function module $\Gamma(F/Y, S)$ is a covariant functor that evaluates to $X@\Gamma(F/Y, S) = \{X@c : X@_S Y \subset X@Y \stackrel{X@C}{\to} S\}$ for functors $C : @Y \to S$ as introduced above. For $h : Z \to X$, we have the map $h@\Gamma(F/Y, S) : Z@\Gamma(F/Y, S) \to X@\Gamma(F/Y, S)$ sending $Z@c : Z@_S Y \subset Z@Y \stackrel{Z@C}{\to} S$ to $Z@c \circ h@Y : X@_S Y \subset X@Y \stackrel{X@C}{\to} S$. Stepping over to the linear dual S-modules $(X@\Gamma(F/Y, S))^*$ we have a presheaf (contravariant functor) $\Gamma(F/Y, S)^*$ for every filter F/Y and commutative ring S.

Let us now prove that there is a canonical natural transformation

$$\Gamma^S_{F/Y} : @_F Y \to \Gamma(F/Y, S)^*$$

with values in the *S-module presheaf* $\Gamma(F/Y, S)^*$ for every filter F/Y and commutative ring S.

Given an object X, the map $X@_F Y \to @X\Gamma(F/Y, S)^*$ is defined as follows. Take an element $f \in X@_F Y$. Take an element $g \in X@\Gamma(F/Y, S)$, i.e., $g : X@_F Y \subset X@Y \stackrel{X@C}{\to} S$. Then we define the image $f^* = X@\Gamma^S_{F/Y}(f)$ by $f^*(g) = g(f)$. We leave it as an easy exercise for the reader to verify that this assignment defines a natural transformation.

We call a filter F/Y *S-modular* if the natural transformation $\Gamma^S_{F/Y}$ is a monomorphism. For a modular filter, the filter functor $@_F Y$ can be embedded in an S-module functor, i.e., it can be "algebraized". It is an important research question to characterize modular filters.

Let us view how the situation of modular filters could occur with particular function configurations. Let us see how a natural transformation $n : @Y \to S$ could be defined such that for two given different morphism $f_1, f_2 : X \to Y$, $n(f_1) \neq n(f_2)$. This would be a condition to ensure that a filter F/Y is modular. One first looks at $V@Y$ such that there is no morphism $V \to X$ or $X \to V$. In this case, one defines $n(g) = C$, a constant in S, for all $g \in V@Y$. Next, take a V such that there is $u : X \to V$. Then for every $g_i \in V@Y$ with $f_i = g_i \circ u$, we set $n(g_i) = C_i$ for two different constants $C_1, C_2 \in S$. This is possible since $g_1 \neq g_2$ here. We set $n(h) = C$ for all other $h \in V@Y$. For a V with $v : V \to X$, we suppose that $f_1 \circ v \neq f_2 \circ v$. It means that v behaves like an epimorphism. This hypothesis is not automatic. But then, we may set $n(f_i \circ v) = C_i$, and $n(w) = C$ for all other $w \in V@Y$. Such a filter would be modular. Our hypothesis is of course not true automatically, but it gives us a first idea about conditions for modular filters. The hypothesis is mainly a function of the relations R imposed upon the basic category $\mathcal{LDC}$. With the minimal relations imposed upon this category (composition of sub-H-jets), the hypothesis evidently holds. This proves the following proposition.

Proposition 5. *Given a commutative ring S, for the minimal relations imposed upon $\mathcal{LDC}$ (composition of sub-H-jets), every filter F/Y is S-modular.*

9.2 Global Filters and a First Čhech Cohomology Theory

Similar to standard constructions of global objects, we can define global filters as follows. We are given a family $(F_i/Y_i)_i$ of filters, also called charts, together with, for every pair $i \neq j$ of indices, subfilters $F_{ij}/Y_i \subset F_i/Y_i$ and equivalences $\phi_{ij} : F_{ij}/Y_i \stackrel{\sim}{\to} F_{ji}/Y_j$ of subfilters. As usual, we ask for compatibility of such equivalences in the following sense: If we are given three "charts" $F_i/Y_i, F_j/Y_j, F_k/Y_k$, together with three subfilter equivalences $\phi_{ij} : F_{ij}/Y_i \stackrel{\sim}{\to} F_{ji}/Y_j, \phi_{jk} : F_{jk}/Y_i \stackrel{\sim}{\to} F_{jk}/Y_k, \phi_{ik} : F_{ik}/Y_i \stackrel{\sim}{\to} F_{ki}/Y_k$, we require that the intersections map as equivalences:

$$\phi_{ij}|_{F_{ij}/Y_i \cap F_{ik}/Y_i} : F_{ij}/Y_i \cap F_{ik}/Y_i \xrightarrow{\sim} F_{ji}/Y_j \cap F_{jk}/Y_j,$$
$$\phi_{jk}|_{F_{ji}/Y_j \cap F_{jk}/Y_j} : F_{ji}/Y_j \cap F_{jk}/Y_j \xrightarrow{\sim} F_{ki}/Y_k \cap F_{kj}/Y_k,$$
$$\phi_{ik}|_{F_{ij}/Y_i \cap F_{ik}/Y_i} : F_{ij}/Y_i \cap F_{ik}/Y_i \xrightarrow{\sim} F_{ki}/Y_k \cap F_{kj}/Y_k$$

and are compatible, i.e.,

$$\phi_{jk}|_{F_{ji}/Y_j \cap F_{jk}/Y_j} \circ \phi_{ij}|_{F_{ij}/Y_i \cap F_{ik}/Y_i} = \phi_{ik}|_{F_{ij}/Y_i \cap F_{ik}/Y_i}.$$

With this information, the *global filter* of this diagram of filters is defined as the diagram's colimit

$$Glob(F_i/Y_i) = \text{colim}_{ij}(F_{ij}/Y_i \subset F_i/Y_i, F_{ij}/Y_i \overset{\phi_{ij}}{\to} F_{ji}/Y_j)$$

As usual in such a context, we have to investigate the classification problem of such global filters, and in particular the question whether there are proper global filters that don't stem from a covering family of a representable functor $@Y$.

It is certainly helpful to this end to consider the cohomology modules of a global filter in the following sense. Given a commutative ring S, for every simplex σ of the covering of $Glob(F_i/Y_i)$ by the images of the charts F_i/Y_i, we have the S-module $\Gamma(F_i/Y_i, S)(\sigma)$ of functions, i.e., functor morphisms $f : \bigcap \sigma \to S$, together with the canonical restriction homomorphism. The functor morphisms f are given on the intersection functors $\bigcap_{F_i/Y_i \in \sigma} Im(F_i/Y_i)$, which are effectively only given with respect to one of their chart functors. But the gluing of charts guarantees that these functors are defined up to natural isomorphisms, which implies that $\Gamma(F_i/Y_i, S)(\sigma)$ are well defined. This being, we have the usual cohomology S-modules $H^k(S, F_i/Y_i)$ that are derived from singular simplices with values in the function modules.

Let us make the cohomology idea more explicit. If we are given a commutative ring S and a "linearization" presheaf V of S-modules, we define a *V-coordinate function* c on F/Y as being the restriction $c : @_F Y \subset @Y \overset{C}{\to} V$ of a functor morphism C into the functor V. See below for examples of such functors. The set $\Gamma(F/Y, V)$ of such V-coordinate functions is evidently an S-module. In the same vein one may consider the covariant approach $(X\backslash F_i)$ of a family of co-fibered functors $f_i : X \to F_i$ and a corresponding family of subfunctors $X_{F_i}@ \subset X@$

There are two trivial coverings for each, the fibered and the co-fibered situation. For the fibered one, we have the identity (Y/Y) which corresponds to the factorization $f = Id_Y \circ f$ of any morphism $f : X \to Y$. We further have the family $(X/Y)_f$ which ranges over all $f : X \to Y$, and which corresponds to the factorization $f = f \circ Id_X$. A similar construction holds for the co-fibered situation.

Let $(F_i/Y)_{i \in I}, (G_j/Y)_{j \in J}$ be two families of filters (not necessarily covering), and suppose that the second is a refinement of the former. This means that there is a map $q : J \to I : j \mapsto q(j)$ of indices such that $G_j/Y \subset F_{q(j)}/Y$. This defines the associated map on covering nerves that maps singular k-simplices $s : \Delta_k \to J$ of (G_j/Y) to singular k-simplices $q(s) = q \circ s : \Delta_k \to I$ of (F_i/Y) such that $\bigcap_{d \in \Delta_k} G_{s(d)}/Y \subset \bigcap_{d \in \Delta_k} F_{q(s(d))}/Y$. Given a singular simplex s, we set $M(F_i/Y,V)(s) = \Gamma(\bigcap_{i \in Im(s)} F_i/Y,V)$, this inclusion defines a S-linear restriction map of function modules

$$(s,q) : M(F_i/Y,V)(q(s)) \to M(G_j/Y,V)(s).$$

We therefore get a linear map of k-cochains

$$C^k(q) : C^k(F_i/Y,V) \to C^k(G_j/Y,V)$$

taking the cochain c to the cochain $C^k(q)(c)$ that on a singular k-simplex $s : \Delta_k \to J$ takes the value

$$C^k(q)(c)(s) = (s,q)c(q(s)).$$

This refinement function induces an S-linear map of the corresponding cohomology modules:

$$H^k(V,q) : H^k(V,F_i/Y) \to H^k(V,G_j/Y),$$

and for a module functor morphism $p : V \to W$, we have the canonical linear map

$$H^k(p) : H^k(V,F_i/Y) \to H^k(W,F_i/Y).$$

An immediate example of a linearization presheaf V is the linearization of a category. Take the functor $V = S\langle Y\rangle$ that evaluates to $X@S\langle Y\rangle = \langle X@Y\rangle_S$, the free S-module over $X@Y$, in fact the classical linearization. Then we have a coordinate function $C : @Y \to S\langle Y\rangle$ sending a morphism $f : X \to Y$ to the basis element $1_S.f$. More generally, for each endomorphism $g : Y \to Y$, we get the evident endomorphism $\langle g\rangle : S\langle Y\rangle \to S\langle Y\rangle$, sending the basis element $1_S.f : X \to Y$ to $1_S.g \circ f$. This entails coordinate functions $C_g := \langle g\rangle \cdot C$. Moreover, for every filter F/Y, we can divide the S-module functor $S\langle Y\rangle$ by the subfunctor $S\langle F\rangle$ generated by the image of F in Y, defining the quotient module functor $S\langle Y/F\rangle := S\langle Y\rangle/S\langle F\rangle$. And more generally, one may divide by a sum of such submodules for a family $(F_i)_i$ of filters: $S\langle Y/(F_i)\rangle := S\langle Y\rangle/\sum_i S\langle F_i\rangle$. Dividing through a subspace $S\langle F - G\rangle$ defined by all differences $f - g, f \in X@F, g \in X@G$ enables a gross identification of filters $F/Y, G/Y$.

The above construction can be evaluated at all arguments X since they all refer to functors. For example, $\Gamma(F/Y,V)$ is a set of functor morphisms, and we may consider the evaluation $X@\Gamma(F/Y,V)$, which is the set of maps $X@q : X@_F Y \to X@V, q \in \Gamma(F/Y,V)$, and which is again an S-module. The same logic allows to evaluate the sets $M(F_i/Y,V)(s)$ yielding S-modules $X@M(F_i/Y,V)(s)$. Therefore we obtain a cohomology theory $X@H^k(V,F_i/Y)$ for every family $(F_i)_i$ of filters and every object X, together with a homomorphism of S-modules

$$H^k(V, F_i/Y) \to X@H^k(V, F_i/Y).$$

What is the musicological content of such a cohomology? To begin with, a cochain $c \in C^k(F_i/Y, S\langle Y\rangle)$ associates with every singular k-simplex $s : \Delta_k \to J$ an element $c(s) = \sum_e \lambda_e.e|Im(s), \lambda_e \in S$, where e runs over all endomorphisms[1] $e : Y \to Y$, and $e|Im(s)$ is the restriction of e to the subfunctor of Y defined by the intersection $\bigcap_{i\in Im(s)} Im(F_i)$. This means that in this situation ($V = S\langle Y\rangle$) a cochain refers to restrictions of endomorphisms of Y to subobjects of Y. But the semiotic of $Y@Y$ is the *self-conception of* Y, and the restrictions are linearized perspectives of such self-conceptions restricted to "parts" of Y.

9.3 A Second Cohomological Approach

The following section presents a technical methodology for gluing together local semiotic instances for the semantic topology to a global configuration.

9.3.1 Hjelmslev-Yoneda Functors

Hjelmslev-Yoneda functors (HY functors for short) are defined to be functors

$$HY : \mathcal{D} \to CAT$$

which associate conceptopoi (objects in CAT) with objects on subcategories $\mathcal{D}$ of $\mathcal{C}$. It is important to have a variability of subcategories $\mathcal{D}$, since we shall encounter situations where a special focus is given to a motivated choice of objects, especially of H-jets.

Let us give a type of examples that may occur quite frequently: Suppose we are given a subcategory $\mathcal{D}$ that has no cycles (except the identities). Then there is a subcategory $\mathcal{D}_0 \subset \mathcal{D}$ of all terminal objects. Suppose we are given all images $HY(Y)$ for $Y \in \mathcal{D}_0$. Then one can define (recursively) $HY(Z) = \mathbf{Limit}_{W\in Z@/}HY(W)$, together with the canonical projections. Another option is to consider the maximal sub-conceptopos of the images $HY(W)$ that is compatible with the morphisms between these images. For example, one could envisage the four H-jets stating that "a category has (a) a subobject classifier", "has (b) all finite limits", "has (c) all finite colimits", "has (d) exponentials". Then the HY functor would associae with these four H-jets the conceptopoi of categories that have subobject classifiers, together with the obvious conservative functors. This would generate a small category with the four arrows $T \to (a), T \to (b), T \to (c), T \to (d)$, and the HY image of T would be the intersection of the four images $HY(a), HY(b), HY(c), HY(d)$, which is the conceptopos $Topoi$.

[1] pay attention to our identification of Y via Yoneda with $@Y$.

9.3.2 Čech Cohomology

So far we are given HY functors on subcategories $\mathcal{D}$. It is of course important to construct a system of such functors and their interaction. To this end, Čech cohomology is the adequate tool, which we want to describe now.

The typical situation is the task to describe the conceptopoi associated with a H-jet X. To this end, we look at semantically open sets, i.e., subcategories $\mathcal{D}$ of a category of $\mathcal{C}$. On each such $\mathcal{D}$, we have the set $HY^{\mathcal{D}}$ of functors ("singular cochains") into CAT. We may now consider for a commutative ring R the free R-module $HY(R, \mathcal{D})$ over $HY^{\mathcal{D}}$.

Next, we consider a family $(\mathcal{D}_i)$ of open sets in $\mathcal{C}$. From this data, we derive the cochain module

$$C(n, R, (\mathcal{D}_i)) = \bigoplus_{i_0, i_1, \ldots i_{n-1}} HY(R, \bigcap_{i_j} \mathcal{D}_{i_j})$$

which ranges over all non-empty intersections $\mathcal{D}_{i_j}$ of n open sets from the given collection. We then have the coboundary homomorphisms

$$d_n : C(n, R, (\mathcal{D}_i)) \to C(n+1, R, (\mathcal{D}_i)),$$

which operate as usual via restriction of functors to higher intersections, together with the usual ± 1 signs, and their linear extension to the cochain modules. And since $d^2 = 0$ by $Im(d_n) \subset Ker(d_{n+1})$, we have the usual cohomology modules

$$H^n(R, (\mathcal{D}_i)) = Ker(d_{n+1})/Im(d_n).$$

And as usual, $H^0(R, (\mathcal{D}_i))$ denotes the global sections of the family $(\mathcal{D}_i)$, i.e., the functors on $\bigcup_i \mathcal{D}_i$.

In particular, if we consider the family of open subsets of $X@/$ for an object X of $\mathcal{C}$, we get the cohomology, which we denote by $H^n(R, X)$.

The entire business here is to understand the information about X, which the mathematical structure projections $H^n(R, X)$ generate. This is a kind of mathematical reflection of the semiotics of X in the framework of general structures, i.e., conceptopoi categories.

9.3.2.1 Extensions of Functors

We are investigating the standard question relating to extensions of functors. More precisely: Suppose that we are given two open sets (which are full subcategories) in the semantic topology, $U \subset V$, and suppose that we are given a functor $F : U \to CAT$. Is there a functor $G : V \to CAT$ that restricts to F? This is of course interesting when we deal with Čech cohomology.

To begin with, observe that limits exist in CAT. Given a functor $F : U \to CAT$, one has to generate an extension G by defining its values for all objects $Y \in V - U$, and then on all morphisms $Z \to Y$ outside U. To begin with, we consider the limit category $G(Y) := \mathbf{Limit}F(Y@/ \cap U)$ within

CAT. This means that we look at the F-image of the diagram defined by $Y@/\cap U$. Its limit $G(Y)$ has all its projection arrows $G(Y) \to F(W)$ for the objects $F(W)$ in $F(Y@/\cap U)$, morphisms which are compatible with the morphisms within $F(Y@/\cap U)$. This extends F to all objects of $V - U$. Next, let us construct morphisms between $G(Z)$ and $G(Y)$, which should bee associated with morphisms $f : Z \to Y$. This means to build a morphism $G(f) : \mathbf{Limit}F(Z@/\cap U) \to \mathbf{Limit}F(Y@/\cap U)$. The universal property of limits asks us to define a diagram of morphisms from $\mathbf{Limit}F(Z@/\cap U)$ to the diagram $F(Y@/\cap U)$. But for f, we have a subdiagram of $Z@/\cap U$ associated with the composition of f and the morphisms $Y \to Y@/\cap U$. This generates a diagram of morphisms as required, i.e., then also a morphism $G(f) : G(Z) \to G(Y)$. The only open question is the functoriality of this assignment, i.e. $G(f \circ h) = G(f) \circ G(h)$ for three objects Y, Z, T outside U. This is left to the reader as a straightforward exercise.

Proposition 6. *The restriction map $V@CAT \to U@CAT$ for open categories $U \subset V$ is surjective.*

This implies that for a covering by two open sets U_1, U_2, $H^0(R, U_1, U_2) \overset{\sim}{\to} RU_1@CAT \oplus Ker(p)$, where $p : RU_2 \to R(U_1 \cap U_2)@CAT$ is the canonical restriction.

10

Semiotic Classification of Creative Strategies

Summary. The following approach is a semiotic one, it presents *creativity as an extension of a given semiotic system*. We shall use the functorial semiotic theory developed in the previous chapters to unfold this idea in a mathematically explicit way.

– Σ –

10.1 The General Method of Creativity

This introduction to creativity theory refers to our book *Musical Creativity* [32]. Our approach to creativity is not a psychological trick nor does it incite a state of ecstasy, possibly boosted by drugs and similar devices. It is a process of discovery and invention that begins with an open question and continues with a run through a sequence of well-defined operational steps. It might not be successful in finding an answer to the question, but we can start over and repeat the processual run until we find an answer. Creativity cannot be guaranteed, but there are good reasons for approaching it based upon a clear strategy.

This approach is a semiotic one, it presents *creativity as an extension of a given semiotic system*. We shall use the functorial semiotic theory developed in the previous chapters to unfold this idea in a mathematically explicit way.

As already sketched in Section 3.7, the process of creativity is composed of the following steps:

1. Exhibiting the open question
2. Identifying the semiotic context
3. Finding the question's critical sign or concept in the semiotic context
4. Identifying the concept's walls
5. Opening the walls and displaying its new perspectives
6. Evaluating the extended walls

G. Mazzola et al., *Functorial Semiotics for Creativity in Music and Mathematics*, Computational Music Science, https://doi.org/10.1007/978-3-030-85190-3_10

Step 1, exhibiting an open question. Creativity should start with the will to find something new that will answer an open question. It has a target.

Step 2 consists of identifying the context of this question. This is a natural requirement: We want to know what is the overall position of the question, since for one individual the question might be open but trivial, while for another it might be a big challenge. Also, we stress that the context is a semiotic one. What does this mean? It means that the context should be one of things, thoughts, and signs that mean something. The open question then would be one that creates new expressions and contents when answered. It would extend the given meaningful context. So creation in our understanding is about creating new contents, not just forms.

Step 3 is the moment of focusing on a specific location in the context, a critical concept or thing where we guess that the open question could be made more precise.

Step 4 is very important. It asks us to identify the concept's walls. Walls are a metaphor for properties, characteristics, and specificities of the concept that circumscribe the concept in a more or less explicit form. It is a delicate task here, since some properties might be so subtle that one is barely capable of recognizing them.

Step 5 asks us to consider these walls and to try to 'soften' and to 'open' them. This means that we ask to what extent these walls are necessary for the critical concept, and whether we could possibly find ways to open them and to recognize new perspectives on the other side of the given walls.

Step 6 is the terminal step. It consists of the evaluation of wall extensions we may have found in step 5, and then the judgment of this evaluation's result. Is it a successful extension of the original critical concept or didn't we find an answer to the open question? If we are successful, everything is OK; otherwise, we have to go back to step 4 and find new walls and new extensions, or even to step 3 and look for new critical concepts.

Remark 1 It is, of course, easy to give an example of a creative strategy that very probably would not be successful. Suppose that an open question in number theory is given. The critical concept could be "number." And the wall one could discover could be "the number's color" when you write it down on paper, on a blackboard, or on a computer screen. You could have the 'ingenious' idea that the number's color could be extended from the usual black, blue, or red to any fancy color. But it is clear that such an extension would not solve the original number-theoretical question.

Remark 2 We should not conclude this informal introduction to our creativity process scheme without pointing out the deep impact of representing ideas and methods in a sensual way (touch, smell, taste, hearing, sight). We have already established the sight- and touch-oriented metaphor of a wall described in steps 4, 5, and 6. When unfolding creative processes, we should always embed our activity in an object-oriented environment, where we feel at home and comfortable. Humans cannot think in abstract categories without using sensual metaphors. Even the most abstract mathematical thoughts are always embodied in objects that can be manipulated by human gestures. This is one of the secrets of successful mathematicians. Einstein, for example, admitted that his thoughts were always performed with an intense chewing activity. He was literally chewing ideas!

Last but not least, we should be aware that probably the most creative force is love for life. After all, this is the strongest motor of human propagation. For a successful creation, the atmosphere should be a warm and loving one[1]; creation out of hate or indifference is never for life. *A creation is a mental baby; it needs a 'mother' and her carrying body.*

10.2 The Three Basic Strategies in Creativity

In the following sections, we will display three basic approaches to creativity. They are defined by the style, which is applied to the process as described above. In concrete cases, often we'll have to envisage a mixture of these three approaches, but nevertheless it turns out to be true that these three directions appear to be fundamental 'vectors' of creative activity.

The previous section dealt with extensions of the semiotic body by creative actions. We now present three types of walls of creative extensions:

(1) *Invalid concepts*, exemplified by Einstein's time revolution, or the strong counterpoint dichotomies.
(2) *The problem-solution duality*, exemplified by Grothendieck's extension of cohomology.
(3) *The labyrinthic search*, exemplified by Wiles' solution of the Fermat conjecture.

All three extensions are generating specific and very different semiotic challenges.

In a first approximation, type (1) is the easiest: it can be dealt with by inspecting the invalid concept's characteristic attributes. This can however be tricky since those attributes might be hidden in a naive approach. For example, in Einstein's time revolution, the singular mode of the invalid time concept is quite hidden to the common sense approach.

Type (2) is less easy since it is unclear how the problem qua semiotic entity could be flipped into a solution, the other side of the coin or Moebius strip. There is an unsolved question here: why is this type of problem a Moebius strip? One should also focus on this type by eliminating the two other types, which is everything but obvious.

Type (3) is the most problematic: To begin with, why isn't it possible/reasonable to try the first two types? For what reason should one run through a

[1] Sidney Lanier says, "Music is love in search of a word."

conceptual labyrinth rather than look for better concepts or two-sided aspects of a problem? This selection among alternative types is by no means exclusive, because a type (3) solution can later be replaced by a type (2) solution, for example. But there is, even when one has decided to stick to this third type, strictly no method(ology) for running through a conceptual labyrinth, except the higher and virtually infinite combinatorics of trajectorial variation.

The next step in our overall endeavor should be an analysis of the semiotic setup for each of the three creative wall types. This is the research direction we are about to elaborate in forthcoming papers. For the time being, we have proposed a functorial semiotics as a conditio sine qua non *for computational creativity.*

10.2.1 Type (1) Walls

This case takes place if and when the semiotic extension is asked for because a deficient, wrong, or inadequate concept is at stake. The semiotic situation is a conflict of two H-jets Y, Z in the sense that the intersection $Y@/ \cap Z@/$ of the two elementary open neighborhoods $Y@/, Z@/$ includes a logical contradiction. This is the easiest type of creative procedures. What nevertheless does not mean that is trivial. Finding the root of a contradiction, the "wall" of creativity, can be very delicate. Let us discuss this with three examples:

10.2.1.1 Albert Einstein's Critique of the Newtonian Time Concept

This example deals with these two H-jets: Y is the reference system transformation in the traditional Galilean form, whereas Z is the H-jet encoding the experiments that show that the light speed is independent of relative speed, together with the non-standard transformation rules for the electromagnetic Maxwell equations. This exhibits a conceptual deficiency, which is solved by a critique of time transformation. The critical concept here is "time" in $Y@/ \cap Z@/$. In Newtonian tradition, the time concept is in a singular form, there is a single time for everybody ("God's global time"). The wall here is the singular of time, which is quite hidden at the onset. Einstein's creative contribution is the thesis that time is not a singular but a plural, one time coordinate for every reference system, and the Lorentz transformation connecting space-time coordinates of two relatively moving reference systems.

10.2.1.2 Cecil Taylor's Critique of the Elementary Components in Jazz Improvisation

In this situation, we are confronted with the following two H-jets: Y is the traditional lead-sheet driven way of jazz improvisation, whereas Z is the jazz pianist's Cecil Taylor embodied approach to improvisation. The conflict here in $Y@/ \cap Z@/$ is the incompatible nature of improvisational approaches when the lead sheet formalism is confronted with Taylor's embodied gestural style. He complains that in Western music, the body played no role. And that this implicitly also holds for the lead sheet approach. This is a critique of the way, in which musical performance, and more specifically: improvisation, is understood. Taylor then proposes a solution where he "imitates a dancer's leaps and movements" with his hands' gestures over the keyboard. This means that the improvisation is no longer defined by a lead sheet score logic, but by the dancing gestures of his hands and body.

10.2.1.3 Counterpoint

This example involves two conflicting instances of classical counterpoint: Y is the concept of a consonant/dissonant interval as inherited from the Pythagorean theory, where small ratios of intervals relate to consonances, such as $2/1$ for the octave, $3/2$ for the fifth, and $4/3$ for the fourth. But in the classical Fux theory, a fourth is dissonant; this is the Z H-jet. This is a contradiction that remains unsolved in the classical approach. In Mazzola, GuerinoMazzola's counterpoint theory, this problem is solved by a new definition of what is a consonant/dissonant interval. The wall against creativity here is the very approach to defining an interval quality. In the traditional approach, an interval is qualified as being consonant/dissonant according to its individual fraction, independently of other interval qualifications. In Mazzola's approach, the qualification of an interval is no longer an individual property, but the property of being a member of a set of intervals, consonant or dissonant. This might be a major wall: the definition of a set replacing the definition of its elements! The sets K, D of consonant or dissonant intervals are defined directly by the requirement of a unique symmetry $A : K \xrightarrow{\sim} D$ that connects the two sets. These sets are defined by the symmetry's existence, and only after the analysis of such symmetries (called "autocomplementary symmetries"), the members of K are defined as being consonant. It turns out that the set $0, 3, 4, 7, 8, 9$ of intervals (0 for the prime, etc., 9 for the major sixth) is a canonical solution, where the fourth 5 is missing, it is dissonant.

This approach not only solves the classical conflict of the dissonant fourth, it also gives a non-psychological explanation of the forbidden parallels of fifths, see [2] for a detailed exposé of this counterpoint theory, where it is also proved that counterpoint can be extended to arbitrary microtonal contexts.

10.2.1.4 Creativity for Denotators and Similar Signs

The first situation: replacing invalid concepts by better ones looks quite feasible for certain types of signs, namely denotators. We have seen in Section 4.7 that denotators are quite transparent signs in that their characteristic content features appear as mathematically precise coordinates, while their signification components are the underlying forms which are, too, transparent in their conceptual architecture. This means that the critical review of a denotator concept can be performed by a trajectory through its different and well-defined components.

When dealing with the concept of time, we may apply a similar critical review; its characteristic contents are

- time $\rightarrow$ physics
- time $\rightarrow$ real number
- time $\rightarrow$ coordinate in inertial frame of reference
- time $\rightarrow$ singular (there is only one time)

A critical review of these attributes could be as follows:

- time $\rightarrow$ physics: replace by physics (no change needed)
- time $\rightarrow$ real number: replace by a *complex* number, this is the approach by Stephen Hawking and Itzak Bars
- time $\rightarrow$ coordinate in inertial frame of reference: this remains valid, but see the next point
- time $\rightarrow$ singular (there is only one time): replace by *plural*, there are as many times as inertial frames of reference; and their transformation is given by the Lorentz transformation

This procedure can be interpreted in terms of the functorial/topological approach: We consider the covariant functor $Y@$ or the elementary neighborhood $Y@/$ of a H-jet Y and review its evaluation $Y@X$ at H-jets X which play the role of generalized contents of Y. The critical review would then have to inspect all these codomains X that are related to Y by morphisms $Y \rightarrow X$. This inspection must be made precise, otherwise the enterprise becomes intractable. In our example of time, the categories of variability of X codomains were very limited: coordinates from real to complex number, or singular grammatical mode to plural mode. Of course, there is no guarantee that the new concept solves the problem, but as a point of departure, one should select very limited codomain environments, following the principle of economy: change as little as possible.

This however presupposed that we have a *topology on our semiotic category* in order to define what are the nearest H-jets for a given one. We have the semantic topology that would offer a first approach. This topic is however quite delicate, as we may illustrate with the above example of the field extension. When we look for the conceptual elementary neighborhood of a polynomial P, it is evident that the polynomial algebra where it is an element shows up. But this means that set-theoretically speaking the polynomial is a content of the algebra. It is not sufficient to look for the content of the polynomial concept P, but also for the expressions for which the polynomial is a content, which means to consider the H-jets contained in the closure of Y in the semantic topology. This is shown in the following morphism graphic:

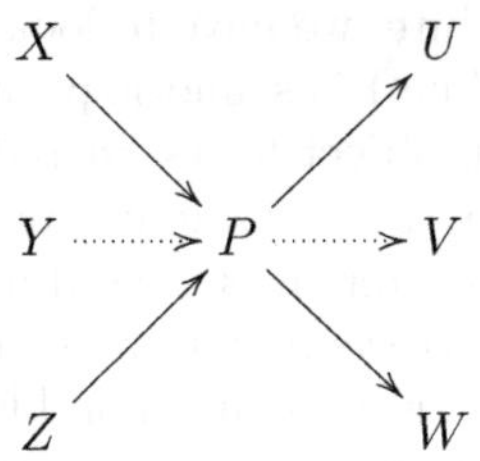

The immediate idea of a topological character is to look for morphisms starting or ending in P, which are composed of at most n arrows of the original directed graph $\mathcal{LD}$. But this is too coarse since even for those morphism that are original arrows in $\mathcal{LD}$, there may be very different relationships to P. For instance, P can be element of the algebra $\mathbb{R}[X]$, but also element of $\mathbb{Z}[X]$ in the case of $P(X) = X^2 + 1$. Or it may be one of the contents of the concept of a Gaussian plane, etc. This seems to relate to metrical distance, not only topology.

Thesis 3 *This problem of a topology/metric on* $\mathcal{LDC}_{\mathrm{R}}^{\Sigma}$ *must be solved to meet the requirement of a mathematical or even computational theory of creativity.*

We believe that searching for alternative H-jets cannot be built upon a random walk, but by the rationale of stepping form a given "defective" concept to a somewhat similar, and hopefully better one. After all, the defective concept is rarely completely wrong since it was successful in previous stages of a development.

What should be required from a concept (H-jet) in terms of topological neighborhoods? Before we embark in this topic, we should observe that the classical idea of a sign's *paradigm* (as opposed to its syntagmatic neighborhood within a concrete juxtaposition of signs) is understood as being its associative field, such as "van", "vehicle", etc. being in the paradigm of "car". This idea is however too vague to be useful here, it reaches from synonymy to similarity and has no operational precision so far.

What is required to define topological neighborhoods is a precise access to variable attributes, such as being in "singular mode" for the time concept. Here the variability, the alternative of a "plural mode", is a precise alternative. If a concept misses precise attributes, its usage in a creative context fails. For example, if the concept "time" has no specification, we cannot embed it within any creative discourse. The specification of a concept, i.e., a H-jet Y, to be precise, would be the reference to all of its arrows where it is in the tail or in the head, as shown in the above graphic. We called the collection of arrows $Y \to Z$ essentially *the elementary neighborhood of* Y, 'dually', the arrows $Z \to Y$ are essentially the points in the closure of Y in the semantic topology, they are the points of the elementary neighborhood of Y for the expressive topology. We denote the set of arrows in the Y closure or in the elementary neighborhood of Y by $Elc(Y)$, its objects are a subset of $@Y@$, call it Y*'s elementary context.*

This means that in a creative procedure, we have to look at all the elements of $Elc(Y)$ and their variability to redefine Y's semiotic position. This is akin to a recursive situation: stepping from an object to its "ingredients". The difference is however twofold: first, the part $@Y$ is not recursive, but 'precursive', such as the contrapuntal situation where an interval is viewed as being an element of the set of consonances. Second, the arrows may have a circular nature, whence recursion would be interminable. This is a technical problem that must be dealt with in a creative formalism that is based upon the reference to $Elc(Y)$.

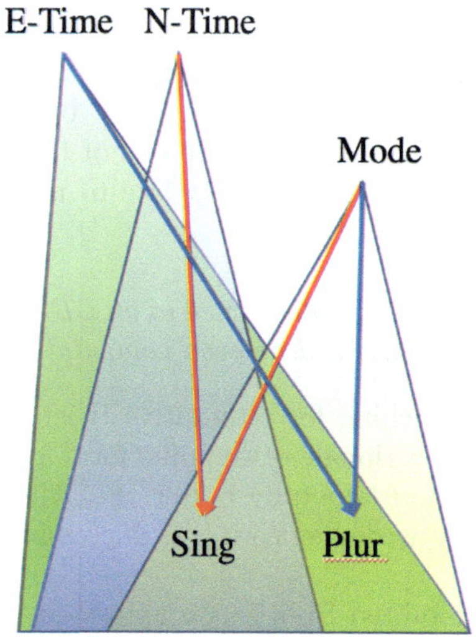

Fig. 10.1: Type (1) topology.

10.2.1.5 Semantic Topology for Type (1) Problems

The situation where creativity works by replacing a wrong concept (type (1)) can be visualized using the elementary neighborhood $Y@/$ of the critical concept, i.e., H-jet Y. The wrong concept Y has some deficient properties. E.g., the singular mode in the Newtonian time concept $N - Time = Y$. This means that in $Y@/$, we have an arrow $s : Y \to Sing$ pointing to the property of being in singular mode $Sing$. This mode is in the elementary neighborhood of $Mode@/$, the concept of grammatical modes: singular ($Sing$) or plural ($Plur$). This means that one is looking at the elementary neighborhood of a concept, which has the wrong property ($Mode \to Sing$ in our example) as one of its semantic values, see Figure 10.1. The type (1) solution comes from the dominating concept of $Sing$, namely $Mode$. Topologically, we look at the elementary neighborhood of a defining concept for the deficient property. We then replace the deficient property by another property in that neighborhood (*Mode@/* in our example).

10.2.2 Type (2) Walls

So far we have dealt with the creative action to be taken if a concept is defective: consider its elementary context and reposition the concept with respect to varied attributive elements of that context. The more challenging action regards the situation we encountered in Galois theory: Here all concepts are functioning, but we have to define new concepts to solve a problem, new numbers, new objects or morphisms of categories, etc. Often such new entities are related to universal properties, such as the adjunction of morphisms that transform given morphisms into invertible ones (see Peter Gabriel's and Michel Zisman's calculus of fractions [11]). However, the calculus of fractions is more generic, i.e., it is not directly building upon the very problem. For example, the extension $\mathbb{N} \subset \mathbb{Z}$, to be described in Section 10.2.2.1, works with the equations $a + x = b$ being the objects of the extension, modulo the well-known equivalence relation among the ordered pairs (a, b) of coefficients. We need a semiotic extension that is not embedded in a larger universal problem, because the universal problem solution is a second step of creativity, generalizing the classical cases to an abstract category-theoretical context.

The extension from $\mathbb{N}$ to $\mathbb{Z}$ and from $\mathbb{Z}$ to $\mathbb{Q}$ are two such situations, which the calculus of fractions covers, but Galois theory is yet another situation. The polynomial equation that needs to be solved is not related to invertibility properties. It is related to the ideal of all polynomials that vanish *together with* the given polynomial. Once this ideal is considered, the passage to the quotient ring is immediate. In both situations, the extension from $\mathbb{N}$ to $\mathbb{Z}$ and from $\mathbb{Z}$ to $\mathbb{Q}$ and those from a field to its extension, the creative action boils down to the creation of objects that represent the problem, sets of equations or vanishing polynomials in our cases, and then the passage from such objects to the corresponding quotient structures. Still another situation happens in the construction of the extension $\mathbb{Q} \subset \mathbb{R}$. A priori it is again a "problem = solution" case: We consider the set $\mathcal{C}$ of not necessarily convergent Cauchy sequences of rational numbers. We then focus on the ideal $\mathcal{O}$ of zero sequences and then define $\mathbb{R} := \mathcal{C}/\mathcal{O}$. Here we don't deal with invertibility criteria as in $\mathbb{N} \subset \mathbb{Z} \subset \mathbb{Q}$, but with convergence, a topological perspective. It is however a sample of the model, where an ideal and an associated quotient structure are at stake.

Regarding the conceptual context of solutions of prominent mathematical problems, the *Weil Conjecture*, solved in 1974 by Pierre Deligne, and the *Fermat Conjecture*, solved in 1994 by Andrew Wiles, the former is a strong consequence of Grothendieck's creative extension of cohomology theories in algebraic geometry, while the latter is a virtuosic masterpiece that does not, to the author's mind, evidence the condition of a creative conceptual extension. Perhaps should one envisage the thesis that for every mathematical problem there should exist a proof that engages in the problem's original conceptual setup and that is built from a creative conceptual extension instead of complex technical procedures.

The situation for type (2) walls boils down to the question whether the very concept of a problem might be misleading, might be the hard part of this wall. If the problem is the front side of the solution, its back side, then the actual conceptualization must be questioned. In other words, can we re-word the problematic sign so as to find its solution-evidencing aspect? We start this discusssion with some examples.

10.2.2.1 Introducing Integers, Rationals, and Real Numbers

Let us first focus on the mathematical examples, the question of defining integers when starting from the natural numbers. The initial question/problem there is that we are faced with potentially unsolvable equations $a + x = b$ for given natural numbers a, b. In these terms an equation is a problem: can it be solved or not. This approach is however not what we are looking fore. We need a concept of an equation *independently* of the potential existence or non-existence of solutions. This means that we have to investigate the semantic H-jet neighborhood $Equation@/$, in more standard terms: the meaning of the word "equation".

In our example, the equation $a+x = b$ means that we are given a domain $\mathbb{N}$ of unknown quantities x and we are evaluating the two sides $a + x$ and b for any given $x \in \mathbb{N}$. This means that we have a map $F(a+x = b) : \mathbb{N} \to \{<, =, >\}$ that evaluates to one of the three relations $a + x < b, a + x = b, a + x > b$. The solution of this equation would be the fiber $F(a + x = b)^{-1}(=)$. But this setup does not presuppose that $F(a + x = b)^{-1}(=) \neq \emptyset$. This definition of the term "equation" works without thinking of solutions. Moreover, for our example, the solutions of this equation, if they exist, are the same iff the maps are, i.e., $F(a + x = b) = F(u + x = v)$ iff their potential solutions coincide, which is equivalent to the well-known equivalence relation $(a, b) \sim (u, v)$ iff $a+v = b+u$. We may therefore identify the equations with their content in terms of the F construction. The next step consists in extending the original arithmetics of addition and multiplication to the F-shape of equations. Evidently, for addition this can be done by setting $F(a+x = b)+F(u+x = v) = F(a+u+x = b+v)$, which is well defined, etc. for multiplication. Therefore the technical approach that replaces numbers by equations qua maps produces the solution by those veritable map equations. Here the problem qua equation becomes the solution, it is the same object, but with another content.

It is a useful exercise for the reader to go through the construction of $\mathbb{Z} \subset \mathbb{Q}$ and $\mathbb{Q} \subset \mathbb{R}$, the rationals and reals in the same spirit as the above construction of integers. For rationals it would be another evident equation, and for the reals it would be the Cauchy sequences to become new numbers.

10.2.2.2 Dodecaphonic Composition

Arnold Schoenberg introduced his method of composition with twelve tones (the dodecaphonic method) in his composition op. 23 in 1921. This was the type (2) solution of the following problem: At the beginning of the 20th century, the classical consonances and their harmonic context became an increasing problem: where would modern composition draw its limits against dissonances with their growing presence. In the first decade of the 20th century, Schoenberg, Anton Webern, Alban Berg, Ferruccio Busoni and others extended compositional contexts to include more and more dissonant harmonies and also new musical forms. Let us call this H-jet Y the "atonal expressionism". The problematic status of this concept was not that it was a wrong concept, it was an inevitable result of musical creativity. Schoenberg's dodecaphonic solution would not eliminate dissonances in the sense of a "restaurative" retro-directed movement, but the dodecaphonic method transmuted the problematic dissonances by "emancipating" them, by integrating them in a new system Z of interval sets, where consonances and dissonances would altogether play a new role that was distinct from the traditional dichotomy consonance/dissonance. Schoenberg's method would suppose a row of all twelve pitch classes, i.e. a sequence $i_1, i_2, \ldots i_{12}$ of all twelve pitch classes, as a germ for a dodecaphonic composition. Interval qualities involved in the setting of such rows and their generically 48 transformations by transpositions, retrogrades, and inversions, were completely irrelevant. The problem of dissonances was turned into the solution by their "emancipation". This method was a compositional perspective that did not deal with aesthesic criteria, the psychological qualities of beauty or ugliness of intervals was not the point of view here, to be clear.

10.2.2.3 Sins and Jesus

The previous examples of type (2) walls were taken from mathematics or music theory. We have no example so far that is not somehow mathematical. This is a serious deficiency. Let us therefore propose an example outside mathematics: the role of Jesus in the Christian religion. According to this religion, humans are born in sin. This system is a (however badly) 'working' environment, nothing will change it. But the problem then remains that humans would love to be reconciled to God. This desire has been fulfilled with the introduction of Jesus, who was sacrificed by God, his Father, and thereby enabled humans to be reconciled to God. The problem was that humans are born in sin without any possible resolution of this deficiency of human existence. The introduction of the (re)solution through Jesus is of type (2): the addition of a divine and simultaneously human instance extends the given (religious) semiotics and implies the problem's solution. It is a procedure that starts with the H-jet's neighborhood $Jesus@/$ that incorporates this Son's life as a human reality, among others, such as miracles, or the immaculate conception. Jesus attracts and absorbs all human sins, is sent to hell (*My God, My God, Why Have You*

Abandoned Me?), and after this divine annihilation generates the Son's resurrection. The problem, our sins, is "divided away" with Jesus absorbing them and giving them, so to speak, back to his Father. Therefore, the problem is converted into its (re)solution.

This example demonstrates that the type (2) wall is well described even in a radically non-mathematical case. But at the same time it becomes evident that the semantic content of this model's components must be stretched to very different directions that have a philosophical/theological rather than mathematical flavor.

Our strategy has two aspects. First: how can we find a generic construction method, and second: how can we know that such a desired object, to be constructed by that method, could exist. It might be that our method leads to a non-existent object, such as a circle with four edges. But here we have to question the criterion of existence, too. In the sciences, existence means not being in any logical conflict with the given system. We are however aiming at a general semiotic, and here, logical consistency of its objects is not a characteristic feature. A unicorn or a circle with four edges exist in a special sense. It excludes them from logical reasoning, but logical consistency is not a global condition for signs.

10.2.3 Some General Ideas for Type (2) Creativity

The general method for type (2) creatvity would be this:

1. Exhibit the conceptual neighborhood (in the semantic topology) for a specific H-jet, qua problem, which we call *conceptual (problem) ideal.*
2. Construct a *quotient (solution) semiotics* that solves the problem, turning the 'problematic' conceptual ideal into the problem's solution.

Before we embark in this methodology, it should be stressed that the principle of the problem being the solution generates a typically creative wall situation. If we argue that the identity "the problem is the solution" holds, the naive duality "problem versus solution" cannot be the ultimately valid understanding of the given configuration of concepts: If that identity or deep nexus is valid, the naive duality must be fundamentally erroneous. The naive duality must be a kind of wall that should be creatively softened and eventually eliminated in favor of a conceptual understanding that dissolves that duality.

Although we don't see a resolution of this conceptual wall, there is a strong positive argument that relates to jazz improvisation. Keith Sawyer in his theory of group flow [40] describes the example of an intrinsic flow that takes place in free jazz improvisation: "In a 60-minute performance of a fully improvised play, the group spends the first half of the performance finding problems they will solve in the second half." This point of view is, of course, totally wrong. Never would John Coltrane or Cecil Taylor or the author Guerino Mazzola of this book construe their creative music in this way. They would rather be engaged in an in-depth exploration of musical universes, but not qua problems, rather in the spirit of creative making *without* question marks whatsoever.[2]

The interesting point in this example is that exactly *because* of Sawyer's erroneous understanding of free improvisation, this art could offer an alternative approach to the naive duality "problem versus solution". After all, the work of free improvisation achieves something, it is not an empty game, neither technically, nor musically. It is not clear at present, in how far the emergence of a flow state could be interpreted in the sense of a conceptual synthesis, a softening of the duality's wall. We would at least expect the co-emergence of an intellectual insight of some kind.

In the following exposition of problem ideal and quotient semiotics, we will not thematize the deep question, why such an approach yields solutions at all. The reason seems mathematically somewhat understandable, but this is a purely formal impression which lacks of deeper rationales. One could argue that the only way to find a solution must start with what is given, but this does not explain how to use the given data, and why a solution is then produced.

10.2.3.1 Abel's General Method

The type (2) situation is also virulent in the strategy used by Andrew Wiles when referring to Yves Hellegouarch's ingenious idea of transferring the (now impossible) Fermat equation $a^n + b^n = c^n$ for $abc \neq 0$ and $n > 2$ to the theory of elliptic curves $E_{a^n,b^n,c^n}(X,Y) = Y^2 - X(X-a^n)(X+b^n) = 0$ (see [16, p. 359 ff.]). Hellegouarch in [16, p. 363] refers to Niel Henrik Abel's "General Method" [1, p. 217]: *On doit donner au problème une forme telle qu'il soit toujours possible de le résoudre, ce qu'on peut toujours faire d'un problème quelconque. Au lieu de demander une relation dont on ne sait pas si elle existe ou non, il faut demander si une telle relation est en effet possible.*[3] This is implicitly the requirement to turn a problem into its solution. The concrete setup for such a method is however not given in its full generality by Abel or Hellegouarch or Wiles.

10.2.3.2 The Idea of a Conceptual Problem Ideal

The conceptual problem ideal is a collection of H-jets that are equivalent to or associated with the given problem H-jet P in the sense that they share the given problem, i.e., generated P is in their elementary neighborhoods. For the

---

[2] It is remarkable that the convergence of scientific problem solving and free jazz improvisation is also thematized by Stephon Alexander in *Jazz of Physics* [3], where Coltrane and Einstein are strongly coupled.

[3] *We need to give a form to the problem which is such that it is always possible to solve it, which can be done with every problem. Instead of asking for a relation when we do not know whether it exists or not, we need to ask if such a relation is possible.*

transition $\mathbb{N} \subset \mathbb{Z}$, this is the set of equations $u + x = v$ that have the same solution x as $a + x = b$ if it exists. For the Galois extension $K \subset K(x)$ it is the set of polynomials $Q(X)$ that vanish together with the given problem polynomial $P(X)$, which means that one takes the ideal $(P(X)) \subset K[X]$. For the extension $\mathbb{Q} \subset \mathbb{R}$ it is the set of Cauchy sequences that converge to the same point—if it exists—with any given Cauchy sequence, i.e., which differ by a zero sequence from this one, defining the ideal $\mathcal{O}$ of zero sequences within $\mathcal{C}$.

The problem H-jet P as such is not entirely arbitrary. It must comply with the condition that it englobes the already given class of signs that we want to extend. It must contain the concept of a 'number' to enable this concept's extension. For our mathematical examples this looks as follows: For $\mathbb{N} \subset \mathbb{Z}$, the equation concept $P : a + x = b$ extends natural numbers as they are solutions of $a + x = b$ for $b \geq a$. For $P : P(X)$, a polynomial over field K, the linear polynomials $aX + b$ have solutions which are the numbers in K. And for the extension $\mathbb{Q} \subset \mathbb{R}$, the constant sequences (a_i), $a_i =$ const., are the rational numbers to be extended. Semiotically set, the H-jet P should have in its semantics a part that identifies with the given class of H-jets.

Let us denote this collection of problem-equivalent H-jets by (P) and call it the *conceptual problem ideal of P*. We may view it as the set/class whose elements are the P-equivalent H-jets. This definition is not precise as it depends on the problem setting when one has to decide what is an equivalent H-jet. It is also not mandatory that this relation to P should be an equivalence relation. But it is plausible that this ideal should be a subset of the totality $T[P]$ of H-jets that can potentially be problematic for the given problem P.

10.2.3.3 Conceiving Quotient Solution Semiotics

Given the conceptual ideal (P) of problem H-jet P, a quotient semiotics must create a system of H-jets that contains as one element the ideal (P), but other collections within $T[P]$ must also generate elements of the quotient semiotics. In our mathematical examples, this could be cosets of the defining ideal or, in the case of the extension $\mathbb{N} \subset \mathbb{Z}$, equivalence classes for other equations, to be built in the same way as the class of equation $a + x = b$. It seems that the quotient structure, which we denote by the mathematically inspired symbol $T[P]/(P)$, is given by an extension of the ideal's structure to other 'problematic' signs. We cannot give a general recipe for its construction yet. But it becomes evident that the creativity wall we are facing is that the quotient is the solution *since* it is made up by problem-related concepts. The critical question here is the following. For $\mathbb{N} \subset \mathbb{Z}$, why is it possible that equations become numbers? For $\mathbb{Q} \subset \mathbb{R}$, why can Cauchy sequences become numbers? For $K \subset K(x)$, why can polynomials become elements of a field? In these examples, as already observed in the previous section, the problematic entities are naturally endowed with the structure of a type of numbers. Is it true that any of such quotient constructions is naturally endowed with a structure that is required by the problem? It seems that the quotient structure is also defined according to the nature of the given domain, i.e., if this domain consists of numbers (of whatever type), for example, then the quotient structure should naturally inherit those numerical attributes. This is a *condition* imposed upon the quotient structure construction!

This entails a condition that already is imposed upon the definition of the conceptual ideal. We have to define it *in such a way* that it enables a quotient structure that extends the given one. The only marvelous detail is that it is possible to define ideals in such a manner. Perhaps is the existence of a solution less marvelous if one takes into account that sometimes problems may look unsolvable, well, not really: a solution may destroy the original setup. For example, suppose that we ask for the extension $\mathbb{Z} \subset \mathbb{Q}$ to be defined for all equations $a \cdot x = b$, also for $a = 0$. Or if we ask the extension $K \subset K(x)$ to be defined by the equation $P(X) = const \neq 0$ for a constant non-zero polynomial. In these cases the ideal 'eats up' the entire given structure. But this is a proper solution despite its degenerate character. This means that the construction of a solution works by its very architecture, independently of the specific setup: There is no case where the solution fails, it just degenerates for "pathological" initial conditions.

10.2.3.4 The Lesson Learned

We learn from this discussion that the conceptual setup includes two components: (A) the conceptual problem ideal, and (B), the quotient solution semiotics. In contrast to type (1), this setup does not yield a satisfactory procedure. The two components don't have to this date revealed a satisfactory reason why a solution can be constructed. The missing link between the problem and its solution is not explicitly recognized from the two components. They are still only formally framing the question of the content of the Moebius front and back side. The geometric content of the Moebius strip is not yet translated into the conceptual content of a resolved duality. The critical decision here is to understand that neither type (1) nor type (3) walls are the right choice or at least: why they are less likely to solve the given problem. If we can provide such arguments, then the type (2) must provide us with the solution *if we are sure that no other wall type exists.* Such a certitude must give us the hidden rationales for the conceptual interpretation of the Moebius strip. Let us assume now that this hypothesis of wall type completeness holds.

The elimination for type (1) walls can be justified by the observation that an invalid concept occurs only for the situation where the problematic concept describes an external reality, a semiotic type of contents that are not internal to the descriptive level, but extend to a wider semiotic framework, such as time in physics. A concept that has only an internal content, such as a mathematical concept, cannot be invalid. There is no evaluation with respect to an independent outside. Except perhaps for those mathematical theories, which turn out to have only on "trivial" object that the cover; this in fact happened to Grothendieck, according to Gabriel's oral report to Mazzola.

The elimination of the type (3) case can be accepted if the conceptual problem is fundamental for the given semiotic framework. This means that one must be assured that a technical trajectory within the given framework cannot produce a conceptual breakthrough. For example, the non-existence of $\sqrt{-1}$ in $\mathbb{R}$ cannot be solved by any technical procedure within $\mathbb{R}$. In such a case, the only solution must be found by a type (2) breakthrough, which in our example is produced by the quotient $\mathbb{C} = \mathbb{R}[X]/(X^2+1)$.

These two rationales are negative: They only tell us why *not* to choose types (1) and (3). And they refer to type (2) only because no other type is supposed to exist, tertium ("quartum" here) non datur. Therefore, there is nothing to be contradicted and nothing to be proved *within* the given semiotics. We however have to make sure that such an extension of the given semiotics exists, and we have to generate a corresponding solution of the problem.

10.2.3.5 A Functorial Approach

We may summarize this idea of a conceptual ideal semiotically as follows:

1. Exhibit the problem concept P in a specific semiotic context K (such as an equation $a + x = b$ for $K = \mathbb{N}$, a polynomial equation $P(x) = 0$ with coefficients in a field K, a convergent series of rational numbers ($K = \mathbb{Q}$) $a_i \to ?$, etc.).
2. Restate a structural description $S(P)$ of P such that it applies to the given context K.
3. Verify that the description $S(P)$ has the elements of K as (trivial) solutions (e.g., for $a \leq b$, the natural numbers solve the equation $a + x = b$, or linear polynomials $X - k$ have the elements $k \in K$ as their roots, or the constant sequences $a_i = a \in \mathbb{Q}$ have their limits in $\mathbb{Q}$). *This is the wall of such a creative project: The structure $S(P)$ is superfluous in K, so why defining it?*
4. Make sure that the structural description includes not only the solutions of the problem in the given context, but also provides us with the context's structure as a group, field, ring, etc.
5. Verify that this structure of K can be extended by $S(P)$ to what we call the quotient context $T[P]/(P)$.
6. Verify whether the quotient $T[P]/(P)$ solves the original problem P.

We may rephrase this project in terms of semiotic objects and relations as follows (the total figure is broken into two parts for reasons of size):

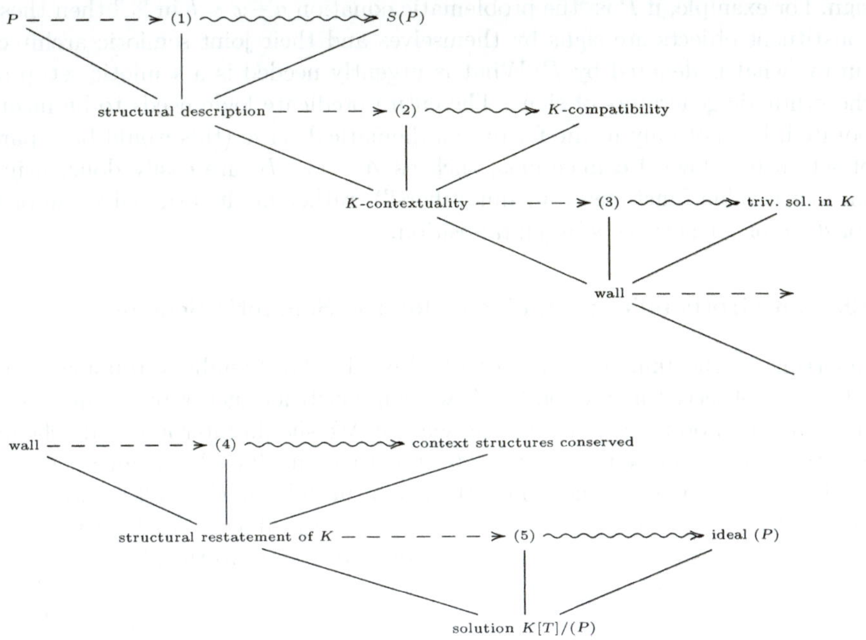

This scheme may then be the semantic template for concrete cases, as shown in Figure 10.2.

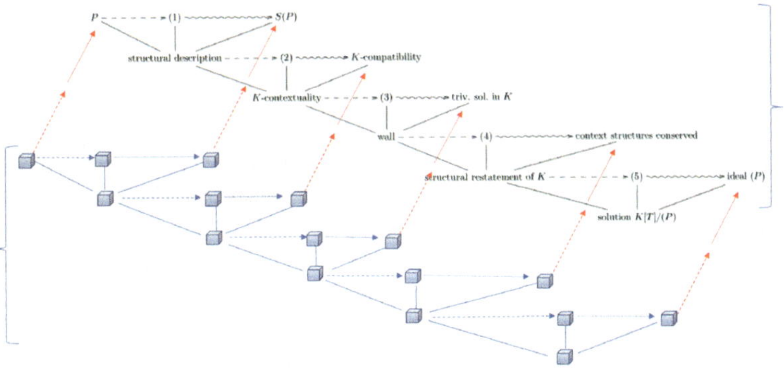

Fig. 10.2: The semantic scheme of type (2) problem solving, together with an instance.

It should however be stressed that the above schematic trees are by no means indecomposable, i.e., P, for example, is not the simple expression of a sign. For example, if P is "the problematic equation $a+x=b$ in $\mathbb{N}$," then these constituent objects are signs by themselves and their joint semiotic architecture is what is denoted by P. What is urgently needed is a semiotic setup of the syntactic generators of signs. The entire predicate logic needs to be incorporated, but not only in the formal mathematical sense (this would be a part of set theory; logical connections, such as $A \implies B$, are easily done, using the truth value functions—but quantifiers?), rather in the original sense of a constructor for new signs by juxtaposition.

10.2.3.6 Grothendieck Topologies for the Semantic Scheme

Referring to the fundamental role of filters for Grothendieck topologies on $\mathcal{LDC}_R^\Sigma$, as observed in Section 3.5.4, we can construct such a topological perspective based on the above semiotic scheme. We should interpret that scheme as a tree S (as discussed in Section 3.6), together with its colimit functor $C(S)$. An instance of that scheme would then be a morphism $F \to C(S)$, where F may be representable or not, but mainly some presheaf associated with part of a tree as shown in the instance part of Figure 10.2. Our Grothendieck topology filter would then be of the form $@_F Y$, which means that one filters through an instance $g : F \to C(S)$ of the creativity scheme $C(S)$. For a given domain H-jet X, we would consider morphisms $f \in X@_F Y$ as points in the "neighborhood" $X@_F Y$, more graphically:

$$\begin{array}{ccccc} f : X & \longrightarrow & F & \longrightarrow & Y \\ & & \big\downarrow{\scriptstyle g} & & \\ & & C(S) & & \end{array}$$

The Grothendieck topology axiom (ii), stability, follows immediately as for a morphism $h : Z \to Y$, the fiberproduct filter $F \times_Y Z$ does the job, and it is projecting into $C(S)$ via $g \circ pr_1 : F \times_Y Z \to F \to C(S)$. Axiom (iii) is not verified here, so we should consider the topology generated by such instances $g : F \to C(S)$.

But what about other cases that are not mathematical, such as the Jesus construction? The above mathematical procedure consisted in giving the term "equation" a specific content that was independent of the existence of a solution. This meant to evaluate the functor *Equation*@ at specifically mathematical locations, viewing the concept as specifying a mathematical map. For the Jesus construction, such an evaluation will not work, but another semantic evaluation should do the job. This suggests that we should in any case

- replace the problematic concept P by its (covariant) functor $P@$ and
- workout its content (which means to evaluate the functor) in such a way as to create the solution by enriching its "system-imminent" potential.

Such a method is strongly connected to the Yoneda philosophy in the sense that objects are replaced by their functors and then embedded in the presheaf category which is always a topos and therefore has a structural richness that the original category may miss. Here, the existence of a representative of a functor is not mandatory, meaning that the problematic nature is suspended. This fact must—by exclusion of anything meaningful to P outside $P@$—be the key to the dissolution of the type (2) wall. Moreover, the specific problem may request non-representable subfunctors of $P@$, which only in the presheaf topos works. On other words, the hope is that the solutions may be constructed from subfunctors of $P@$ without any reference to representing objects. This idea means that rephrasing the problem in a functorial setup eliminates its problematic aspect. In our mathematical examples, this meant to rephrase the concept of an equation without specifying the question of the existence of solutions.

10.2.3.7 Expressive Topology for Type (2) Problems

The topology for type (2) problems is shown in Figure 10.3. Here, we have two

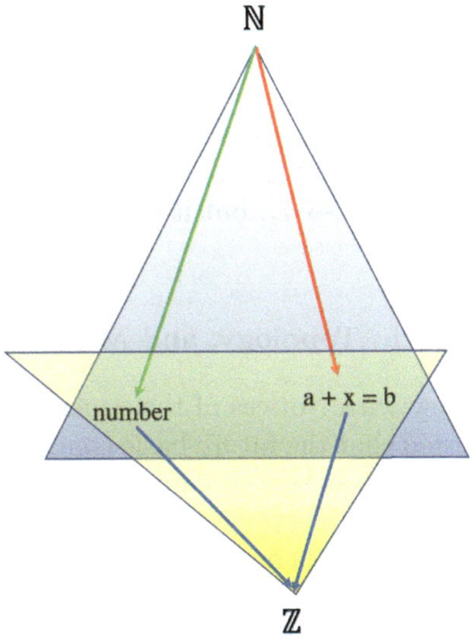

Fig. 10.3: Type (2) topology.

concepts in the elementary semantic neighborhood $Y@/$ of concept Y, in our example the natural numbers $\mathbb{N}$. The two concepts are *number* and $a + x = b$, an equation. The type (2) solution consists of the synthesis of the two concepts, in our example this means that the problematic equation becomes a number, in other words, it solves the problem by introducing new numbers. Semiotically speaking this amounts to the exhibition of a new concept that unites the two concepts in the problematic elementary neighborhood (of $\mathbb{N}$ here). Topologically speaking we have to find a new concept, which is a common content of the two given ones, in other words, the new concept ($\mathbb{Z}$ in our example) contains the two given concepts in its expressive elementary neighborhood $/@\mathbb{Z}$.

There is a fundamental difference between the semiotic situations of type (1) and (2) beyond the topological similarity. And this relates to the existence question. For type (1), we are looking for given H-jets that may generalize a property of a H-jet. In type (2), the two given H-jets are thought of being expressions of a *new* H-jet. This one needn't exist yet, we are asked to construe it to comply with the structural requirement. For example, in our above example, the H-jet $\mathbb{Z}$ must be built from the two H-jets *number* and $a + x = b$. The question here is about *a priori construction methods.*

10.2.3.8 Limited H-jets

A typical solution of type 2 can be found as follows. We are given a H-jet Y of limited type. For example a number domain of type commutative monoid. This H-jet would be the limit of properties "addtitive monoid", specified by their properties of being an associative operation with neutral element, etc. This is defined by a set of arrows $Y \to Y_\iota$, which describe the monoid properties. For the type 2 solution, we would have to specify arrows that correspond to these properties, but now starting from the domain of an equation E. The solution would then be given by defining arrows from E that correspond to the Y-arrows to the Y_ι. The H-jet E would not necessarily be the limit of the arrows corresponding to the $Y \to Y_\iota$, but it would give a starting setup for E's definition as a monoid.

10.2.3.9 Content Search, Topology, and Manin's Suggestion

In a remarkable interview with Notices of the American Mathematical Society [22], Yuri Manin suggested that the future basic concepts in mathematics might not be abstract entities, such as sets or categories, but "sensual objects". This implies that he conjectures that abstraction would not be the ultimate approach to mathematics anymore, but concrete, human objects.

In our semiotic perspective, this amounts to searching for contents of mathematical entities, problems, questions, which would reveal aspects that open up a different understanding of what is looked at. A first step towards this plan are two now successful approaches, namely Gabriel's quiver algebras and algebraic topology. In both cases, abstract algebraic entities are given a geometric/sensual interpretation. Quivers are directed graphs, and algebraic topology enables a "tangible" interpretation of abstract algebra, for example viewing groups as fundamental groups, which reify algebraic elements as classes of curves in topological spaces. For example, the fundamental theorem of algebra can be proven in terms of continuous curves and their cycling around zero.

The semiotic consequence of this approach boils down to search for semantic enrichments of abstract mathematical situations. Such enrichments would include Manin's, Gabriel's and similar directions. But not only these ones, also other contents that could reveal hidden meanings of abstractions. These hidden layers of concepts would ideally reveal access to solutions of problems which were hidden in the original setup. Or, to put it in the framework of our solution types: these layers would enable a type 2 solution, or at least, a solution that would become visible from the semantic enrichments. This is not only about visualization (the topological perspective), but also about sonification (the musical perspective) or a haptic/gestural concretization. Manin's suggestion then would also comprise sensual basic objects of musical and other artistic categories. The strong type 2 conjecture would mean that one should succeed in finding semantic enrichments capable of solving problems in the sense of type 2.

10.2.3.10 Grothendieck's Coconut Metaphor and Wiles' Solution

The above strategy looks particularly dramatic when we recall the nature of Wiles' solution of Fermat's conjecture. The original formal setup $a^n + b^n = c^n$ was transformed into a set of coefficients of a special elliptic curve. The latter has no visible meaning with respect to the original formalism. The transformation of formalisms has no visible rationale. This is a huge problem behind Wiles' solution: It is a typical "trick" in Grothendieck's sense.

10.2.4 Type (3) Walls

So far, we have considered two types of creativity walls: (1) the walls of invalid concepts, and (2) the walls of solutions being the 'hidden agenda' of the problem, i.e., where "the problem is the solution". Einstein's revision of the concept of time is an example of type (1), Deligne's solution of the Weil conjectures is of type (2) insofar as Grothendieck's conceptual extension of cohomology theory is not revising an invalid concept of cohomology theory, but adds new "motivic variations" thereof.

Wiles's solution of the Fermat conjecture is a third type that we have mentioned, but not discussed so far. We have only observed that the solution is the virtuosic achievement built upon a highly sophisticated mathematical technique. Of course, Deligne's work on the Weil conjectures is also, beyond the conceptual extension, highly virtuosic, he earned the Fields medal 1978 for this achievment. In some sense, this type of creativity is similar to the creativity achieved by free jazz creators John Coltrane and Cecil Taylor: These musicians have developed an extremely sophisticated and innovative instrumental technique on their instruments beyond their dramatic extension of conceptual aspects of jazz improvisation and composition.

In this third type of a creativity wall, there are no invalid concepts, and the problem is not the solution, at least if the given conceptual framework is referred to. Such a wall is eliminated by playing around with the box's contents until the box literally explodes. The potential solution is not guaranteed here, the problem may be undecidable in the given axiomatic framework. The solution might be accessible through a complex labyrinth of logical inferences that connect a number of auxiliary results or methods. This became clear also from Deligne's remark, during his presentation of the solution at the IHES in 1974, about one of the "tricks" he applied in his proof, and which Grothendieck did not appreciate at all because it did not fit into his architecture of smart and elegant concepts.

This type of creatively opening the given problem's box (by a veritable 'explosion') seems to lack systematic aspects. One may be playing around in that labyrinth until a path from the problem to its solution is found—including the case of an undecidable problem.

It is especially dramatic in Wiles's solution of the Fermat problem that the trajectory of the solution, as beautifully described by Yves Hellegouarch in [16], shows the crucial moment, namely when (around 1965-1970) Hellegouarch and Wadim Andrejewitsch Demjanenko discovered (seemingly independently) that the existence of points of order $2p^2$ in special elliptic curves implies the existence of a non-trivial solution of the Fermat equation $a^p + b^p = c^p$ for prime number $p > 3$. In 1969, Hellegouarch reversed that observation conjecturing that a non-trivial solution of the above Fermat equation would imply the existence of points of order p in the elliptic curve $Y^2 = X(X - a^p)(X + b^p)$. Such a curve was later wrongly called "Frey curve" following Gerhard Frey's important conjecture in 1985 that such elliptic curves could not verify the conjecture of Taniyama-Weil regarding the modularity of elliptic curves. But the decisive transition from Fermat's formula to elliptic curves remains arcane to this date. There is no evident structural connection between elliptic curves and Fermat's formula. Hellegouarch simply states "À notre grand étonnement l'existence de ces points entrâinait celle d'une solution non triviale de l'équation de Fermat d'exposant *p*." This transition is a veritable "trick" in Grothendieck's understanding. Unfortunately, Hellegouarch does not specify any conceptual connection, the facts are simply "astonishing"—not satisfying to the author.[4] It is one of the many miracles in mathematics, where seemingly unrelated structures reveal creative connections.

If type (3) can be replaced by type (2), one should restart the entire Fermat problem investigation by the original statement, namely, that the subscheme of $\mathbb{A}^3$ defined by the closed condition $X^p + Y^p - Z^p = 0$ and the open condition $XYZ \neq 0$ has no rational points. The big question here is whether there is a type (2) wall whose dissolution would solve the problem by a conceptual moebius strip. This is of course only reasonable if the wall-type principle "every type (3) wall can be replaced by a type (2) wall" is (accepted and) valid.

[4] This mystery remains unsolved even after our recent (August 2017) email exchange.

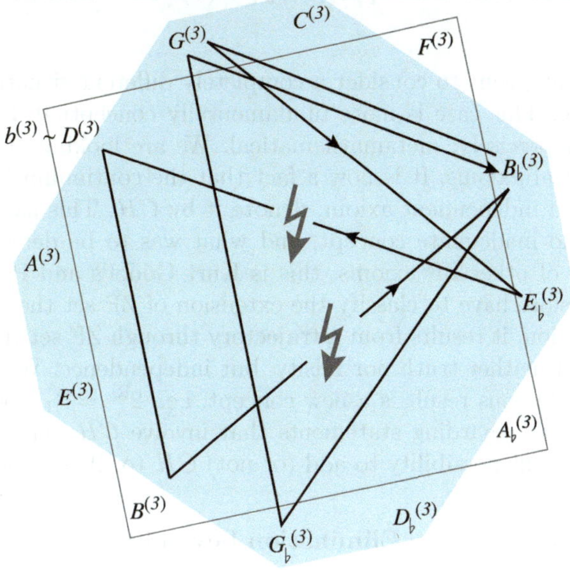

Fig. 10.4: The graph shows the modulation plan in the Allegro movement of Beethoven's op.106 in the tonality system arranged on the circle of fourths. The start switches from $B_\flat$ to G. The inverse modulation occurs at the end, and both follow the same procedure. Except these initial and terminal movements, the modulation plan is perfectly symmetric around the symmetry axis between $B_\flat^{(3)}/E_\flat^{(3)}$ and $A^{(3)}/E^{(3)}$.

10.2.4.1 Ludwig van Beethoven's Type (3) Creativity in the Sonata *Hammerklavier* op. 106, Allegro

In the famous *Hammerklavier* Sonata, Beethoven displays the abstract duality of a "world" around $B\flat$ versus an "antiworld" B, which he unfolds in a complex architecture of modulations between "world" and "antiworld" tonalitiels, see Figure 10.4.

10.2.4.2 John Coltrane's Type (3) Creativity

Coltrane's approach to creativity is a typical type (3) method. He starts with a simple abstract question concerning a tonal setup and then delves into its infinite ramifications, which he performs until all possibilities have been transgressed. In *Giant Steps* for example, the initial setup is an inversion I_D at D, the inner symmetry of C-major. He then unfolds this symmetry in a complex display of a system of chords that are related to each other under I_D. In his more free jazz driven performances, for example in his CD *Coltrane at village Vanguard Again*, he improvises an explosion of ramifications until "everything has been said".

10.2.4.3 A Different Example of Type (3): The Continuum Hypothesis

It seems advantageous to consider a completely different situation where type (3) is at stake. This case is more fundamentally conceptual, but still mathematical, more precisely: metamathematical. We are looking at ZF set theory with its classical axioms. It is now a fact that the continuum hypothesis, i.e., $2^{\aleph_0} = \aleph_1$, is an independent axiom, denote it by CH. This sign is fundamental, there is no inadequate concept, and what was to be demonstrated is its independence of other ZF axioms, this is Kurt Gödel's and Paul Cohen's result. We therefore have to classify the extension of ZF set theory by CH as a type (3) situation, it results from a trajectory through ZF set theory, resulting in the proof of neither truth nor falsity, but independence. What is added by CH after this famous result is a new concept, i.e., $2^{\aleph_0} = \aleph_1$, together with all the consequences regarding statements that involve CH. The solution is the logically consistent possibility to add (or not) CH to ZF set theory.

10.2.4.4 Type (3) Walls: Elimination beyond Combinatorial Efforts?

So far, type (3) walls don't offer any ingenious method that transcends pure combinatoriality (which can be extremely demanding!). This case is all the more dramatic as all three options are at stake: confirmation of a conjecture, finding a counterexample, or undecidability. If one does not stick to the thesis that every type (3) wall should be solvable by a type (2) procedure, there is no visible simplification here.

The typical type (3) case is the problem of angle trisection solved by Galois. This is however solved by Galois theory, which is a completely type (2) construct. Could one therefore ask to reduce a type (3) problem P to a sequence $P_1, \ldots P_m$ of type (2) problems? This is evidently not the case in the Fermat-Wiles solution: The connection of the Fermat equation to elliptic curves discovered by Hellegouarch is a kind of Einstein-Rosen bridge or wormhole in the semiotic manifold of knowledge, see Figure 10.5.

Our question/conjecture is that any such type (3) wormhole can be circumnavigated by a sequence of "ordinary" type (2) steps, see Figure 10.6.

The sequence of type (2) steps is a delicate situation. Often such a step might seem trivial and not solving any relevant problem. Let us look at two examples: First, recall that Galois theory is based on the trivial type (2) step for finding solutions of polynomial equations, that construction of the quotient $K[T]/(P)$. But once this architecture is built, the Galois correspondence becomes an obvious and powerful tool to tackle the classical geometric problems: cube doubling, or angle trisection. Second example: Calculus, especially the definition of instantaneous velocity that was not given when Galilei made his experiments. The basic concept of calculus is the construction of limits of numbers. But this concept was impossible without the introduction of real numbers. The *definition* of limits was only possible in $\mathbb{R}$, which in turn is a simple type

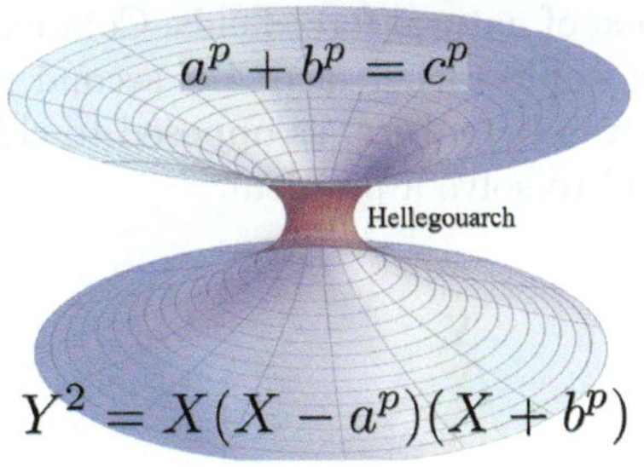

Fig. 10.5: The connection of the Fermat equation to elliptic curves discovered by Hellegouarch is a kind of Einstein-Rosen bridge or wormhole in the semiotic manifold of knowledge.

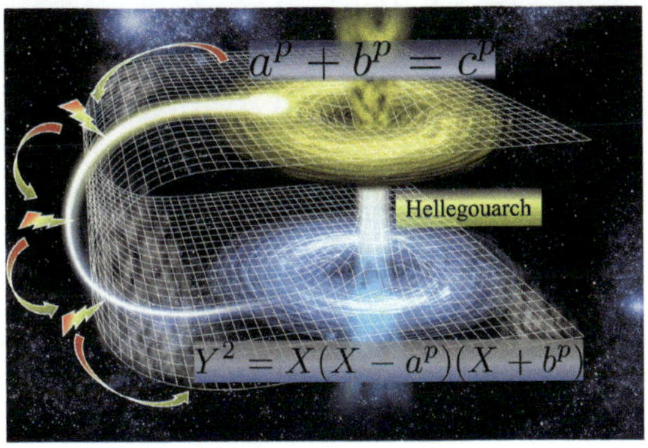

Fig. 10.6: Can any type (3) wormhole be circumnavigated by a sequence of "ordinary" type (2) steps?

(2) extension of $\mathbb{Q}$. What follows from these examples is that type (2) solutions may seem irrelevant for given difficult problems, but they may help solve—and in fact have done so—difficult problems, which would have been inaccessible without the architecture provided by the solution of typ (2) problems.

Conjecture 1. Type (2) steps are the infinitesimal information movements towards type (3) solutions.

This conjecture should be viewed from the conceptual perspective of dynamic systems: We need procedures for the creation of type (2) extensions of P, "tangential arrows" from P, and also concepts of vector field type to represent force fields for semantic selections of such type (2) extensions. The question is of course about the *topology for arrows* and the genealogy of such *semiotic force fields*. While at the present time Grothendieck topologies seem the most natural ones, the genealogy of semiotic force fields is not clear. There are two main threads to be discussed: Either the problem acts as an attractor or the force fields are "naturally' given such that the solution of a problem would result from the integration of natural force fields. Coming back to Grothendieck's coconut parabola, one would rather favor the second thread: warm water and the sun's energy are given by nature, as opposed to chisel and hammer, which are enforced by our will to solve a problem.

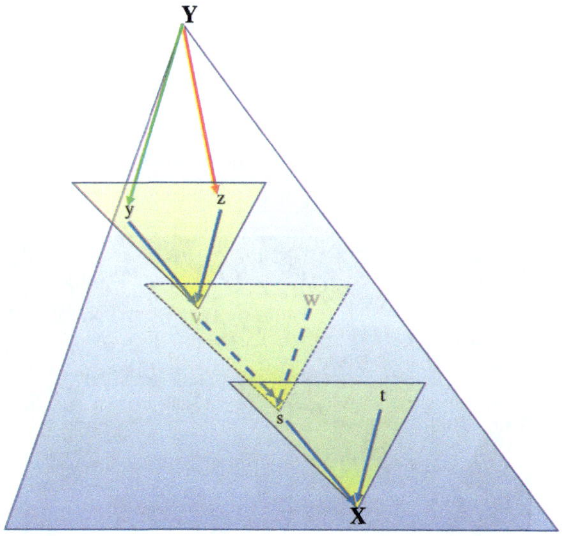

Fig. 10.7: Type (3) topology.

10.2.4.5 Expressive Topology for Type (3) Problems

The topology for type (3) problems, if one accepts the above conjecture 1, is shown in Figure 10.7. We have a sequence of type (2) transitions, which add up to the solution of a type (3) situation.

10.2.4.6 Grothendieck's Credo

This brings us back to the fundamental question about the time-dependency of semiotics. Grothendieck's natural approach presupposes that everything is already written down in an atemporal platonic or otherwise divine sky, and that we simply have to "listen" to this layer of existence. The problem here is the accessibility to those heavenly entities. In practice, we will have to provide tools for the construction of H-jets, nobody will deliver them for free—revelation is not a scientific option anyway. Therefore the creation/construction of H-jets is vital to the approach to problem solving for probably any type (1), (2), or (3).

In other words, the Grothendieck topologies must be understood as devices that go in parallel with the creation of their filters. There are no given/visible neighborhoods, one has to build them before deciding where to go. *Grothendieck's listening metaphor is a petitio principii.* What is needed is an experimental semiotics that tries out H-jet suggestions, and perhaps should be termed "virtual semiotics", where one plays around with sketches/options of H-jets. There is no obstruction against such constructs as long as they don't destroy the semiotic architecture.

Grothendieck had two main principles: A. "Ask the right question!" (well, better: "You are asking the wrong question.") B. Don't open the coconut by chisel and hammer, but by softening its hard surface in warm water under the sun."

They are related to each other. But let us first look at the first, A. It of course lives from the hypothesis that Grothendieck knew or was supposed to know the right question. And then, what would demonstrate this fact? Well, the success. Which is a circular argument. How did Grothendieck know about the right question? Divine revelation? Basic mathematical instinct? This is an unsolved question.

Then, even if one poses the right question, the solution would be enforced by the soft method. But how does one know which warm water to choose? And couldn't this conceptual softening procedure be waisted to a wrong coconut? Again, coming back to point A, there is no guarantee that the question, even if it is the right one, is being solved by the right coconut opening, well: the right coconut (which needn't coincide with the right question!), and the right opening procedure.

This summary of Grothendieck's credo leaves us with some embarrassment. These two principles, A an B, are circular and utterly vague despite their inspiring suggestivity. We could however suggest a solution of this problem by the type (3) to type (2) switch proposed above in Conjecture 1: *Grothendieck followed this conjecture, namely that every type* (3) *solution can be replaced by a sequence of type* (2) *solutions.*

10.2.4.7 Type (3) and Length of Proofs

The type (3) situation is not only related to a number of steps, which are combinatorial in nature, it is also related to the question about the role of the number of steps as such. In other words, the length of the type (3) step number is also a problem when it reaches beyond human performance. For example, the four-color theorem has been demonstrated by a proof that cannot be verified by humans, the proof that was given in 1976 by Appel and Haken [4] gets off the ground by an ensemble of 1500 charts that imply corresponding demonstration, which add up to more than million pages, a quantity that cannot be performed by humans. Computers are needed to transgress all these proof sequences. In his book *Les métamorphoses du calcul* [9], the logician Gilles Dowek argues that such a situation changes mathematics to a science that needs new machines, similar to physics and astronomy, where machines are mandatory as research tools. The title of his book refers to "calcul" as a characteristic feature of mathematics. This perspective is revelatory for an understanding of mathematics as a methodology of calculations. The underlying hypothesis is that mathematics is reduced to calculations, proofs are more or less complicated calculations, and the theory of proofs views proofs as irreducible combinations of calculations. A mathematics of calculations is conceived in this perspective.

So let us first discuss the idea that proof length may generate a limit of human control in mathematics. The question here boils down the problem of finding proofs that are at the reach of human performance. The Appel-Haken proof seems to be an example of a proof (type (3), of course) that reaches beyond human capacities. Can we contend that any such proof may be replaced by a proof that can be controlled by humans? In other words: can we argue that a proof that exceeds human control may for good reasons be a suboptimal proof? Such reasons may be a lack of intelligence, a lack of conceptual control over the problem's architecture. While for the four-color theorem, this cannot be verified, other cases are good examples thereof.

Let us take the problem of calculation $S(n) = \sum_{i=1}^{10^n} i$. The naive calculation of this sum is a human endeavor for, say, $n \leq 5$. But for $n = 9$, the calculation of the sum $1+2+3+\ldots 10^9$ becomes very difficult. And for $n = 15$, say, a direct calculation becomes impossible. Suppose you need one second for one addition, then the calculation needs 10^{15} seconds, roughly 3×10^7 years. Therefore, the problem of calculating $S(n)$ in the naive style is beyond human capacities for large n. Of course, this exceeding complexity can easily be managed by Gauss' famous idea: Instead of adding natural numbers in their natural order it is more intelligent to add the first and the last, yielding $1 + 10^n$, then the second and the one but last: $2 + (10^n - 1)$, again yielding $1 + 10^n$, and so on, until $10^n/2 + (10^n/2 + 1)$, which again yields $1 + 10^n$, this adds up to $10^n/2 \times (1 + 10^n)$. The general formula therefore is $S(n) = 10^n \times (10^n + 1)/2$. This number can be calculated in seconds, and we are back in the human performance.

From this example we learn that a humanly too hard problem can be solved by an extra effort of intelligent insight, i.e., that the negative outcome can be attributed to a too low insight, and not to an intrinsically unsurmountable difficulty. This may imply that the overlong proof of Appel-Haken could be shortened by a more in-depth understanding of the four-color problem. Our general thesis would then state that *a type (3) problem can be solved by a humanly performable approach if our intelligence of the problem's hidden conceptual architecture is high enough.* This thesis should be valid for a sequence of type (2) or type (3) steps.

This thesis opposes Dowek's understanding of mathematics as a way of calculating to the understanding of mathematics as a conceptual endeavor.

10.3 In Search of A Global Geometric Perspective

If one accepts the classification of walls with the three types (1), (2), and (3), and if one sticks to the thesis—which underlies also Grothendieck's credo—that eventually, type (3) can be replaced by a type (2) wall, then the logically stringent next step consists of the search for a fundamental geometry of concepts that could reveal something like the "back" of the Moebius strip in the sense of a hidden solution. In view of the semantic nature of any creative action, such a geometry must emerge from a geometric construction that is related to the semiotic categories introduced and discussed earlier in this paper. See our cohomological setup in Section 9.

Example 10. A simple example: A Moebius Strip Nerve. We want to construct a covering of a final object $Y = 1$ that has a Moebius strip as its nerve. To this end, we take five copies A, B, C, D, E of 1 and define five functors $F_1 = @A \sqcup @D \sqcup @E, F_2 = @A \sqcup @B \sqcup @E, F_3 = @A \sqcup @B \sqcup @C, F_4 = @B \sqcup @C \sqcup @D, F_5 = @C \sqcup @D \sqcup @E$. The covering of the set $\{A, B, C, D, E\}$ by these five subsets has a Moebius strip as its nerve. Now, take the covering $F_1/1, F_2/1, F_3/1, F_4/1, F_5/1$ of $@1$. We then get a Moebius strip as desired.

10.3.1 A Classical Type (3) to Type (2) Switch

The classical theorem of Euclid stating that there is an infinity of prime numbers is a simple type (3) argument: Suppose there are only finitely many primes $2 = p_1, 3 = p_2, \ldots p_N$, then take their product $\prod_{i=1,2,\ldots N} p_i$ and add 1. Then the number $1 + \prod_{i=1,2,\ldots N} p_i$ is not divisible by any of the given primes, therefore there are more than N primes, contradicting the finiteness assumption. The type (2) argument backing this proof has nothing to do with infinity. It runs as follows: For any natural number n take the product $p(n)$ of all primes $p \leq n$. Then $p(n)+1$ has only prime factors greater than n. This defines a map $q : \mathbb{N} \to P$, where P is the set of finite sets of primes. It associates with n the prime factors of $p(n)+1$. All primes in $q(n)$ are greater than n. This is a priori not a statement about infinity. This is the hidden type (2) structure. But it immediately implies that there are arbitrary large primes, therefore their set is infinite. The difference in arguments is subtle, but significant.

10.3.2 Galois' Miracle

The substantial contribution of Galois is the group-theoretical analysis of the conceptual extension of fields. The group theory comes in naturally as an analysis of the inner symmetries of the conceptual extension. A more elaborate discussion of Galois' conceptual ideas would be important. Galois' theory is to our knowledge the first sample of conceptual mathematics (which succeeds the object-focused and structural mathematics).

The case of Galois theory is also important when we ask a type (3) problem to be solved by a type (2) construction. Originally, the problems of angle trisection and cube doubling were hard type (3) problems. The Galois method however essentially transformed them into type (2) problems by the theory of field extensions and their group correspondence. This is a tricky transformation. It could not be anticipated that the type (2) construction of (algebraic) field extensions would imply the solution of those classical problems. In the spirit of Grothendieck (that in fact also worked for the Weil conjectures), one would argue that every type (3) conjecture can be solved by an adequate type (2) construction. The real problem of such a meta-conjecture is that in its general setup, no concrete/effective procedure seems feasible to find such a type transformation. And in the case of the Fermat theorem we don't see that transformation being the case. At present, Hellegouarch's wormhole (see Section 10.2.4.4) cannot be circumnavigated by a type (2) trajectory.

10.4 The Deep Mathematical Architecture: Objects, Structures, Concepts

The deep architectural background of Galois theory relates to the historical unfolding of mathematics, which we view as follows. We recognize three stages of its development.

The *first stage* is a mathematics of objects. It focuses on the analysis of specific objects, such as the real numbers, function spaces, etc. We would position Riemann's achievement as typically being focused on the understanding of such specific objects. The completion of this level can be recognized in set theory, the *mathematics of objects*. This theory provides us with a general machinery for the construction of any classical object. All classical objects can be constructed as specific sets following the ZF axioms. Set theory is a systematic machinery for the construction of single objects, it is an object factory.

The *second stage* would be the mathematical theory of structures, which is an abstraction from the specific objects that aims at describing the structures which can be generated in the sense of specific types of objects, such as groups, rings, digraphs, etc. The theory of such structures is what has been achieved in Bourbaki's approach and completed by category theory. The latter is a *theory of structures* independently of their set-theoretical reifications. Within category theory, topos theory is a synthesis between the thesis of sets and the antithesis of structures: in a topos, set theory is simulated as a structural theory without recurring to the set-theoretical element relation. It is only logical that Cohen's continuum hypothesis proof (forcing) was given a topos-theoretical interpretation. It would also be important to prove a topos-theoretical restatement of the recursion theorem (construction by recursion).

The *third stage* would be what we coin *mathematics of concepts*. It is the theory of mathematical concepts which may be reified in structures[5]. This approach deals with the construction principles and techniques of concepts. Two prominent types of this approach are (1) the construction of objects by recursion and (2) the construction of objects in Galois theory. The task here is the construction of concepts which realize a specific type of conceptual extension. This third type is strongly tied to what we have discussed in the type (2) problem solving. Here, objects are constructed that solve problems of type (2) qua new concepts, they are specified as conceptual extension rather than specific objects or structures. This stage is still embryonal, its mathematical mechanism is not systematically defined yet. It is this type of mathematics that must be investigated in a future approach. It may feel much like category theory that was misunderstood as abstract nonsense, namely being the search for solutions of problems which are the backstage of their solutions. What could be the profit of such a seemingly tautological enterprise?

In the case of category theory, the thought error was to confuse essentialism with abstraction: Grothendieck was *not* in search of abstraction per se, but of the essential component in the problem's structure. Abstraction meant to take away (ab-trahere in Latin) the contingent, those properties which obscure the substance.

In the same spirit, doing conceptual mathematics with type (2) solutions does not plainly restate the problem as solution, but reveals its conceptual potential which enables analytical procedures that were unthinkable beforehand.

In the case of Galois theory, the extension of a field defined to generate zeros of polynomials, a type (2) construction, enables the display of groups of automorphisms that opens the access to a classification of such extensions. It essentially conceptualizes the search for zeros in the framework of group theory. This type of subtleties is purely conceptual, neither of object type, nor structurally focused.

The question is now what type of mathematical setup can be taken as a representative of conceptual construction. We propose that this takes the form of an extension $V \subset V'$, which is an object extension (subset relation), which is also a structural extension and a conceptual extension in the sense that there is a problematic concept in V which is resolved within V'. We have described this with the quotient concept $V' = V[T]/(P)$. The critical point here is the *idea of a problematic concept* P. This has to be made precise, it is the critical issue.

[5] The title "conceptual mathematics" was given to Riemann's style of creating mathematical concepts, in particular: manifolds, as spatial concepts, in [18]. This name relates to concepts that are built as constituents of the mathematical language. Our approach here deals with conceptual extension as a properly mathematical technique, not as a linguistic device. The classical homonymous monograph by William Lawvere [19] is not really about concepts, but about category theory, which is a mathematics of structures, not of concepts.

To this end, one must elaborate a statement that defines the "universal property" of $V' = V[T]/(P)$. This is not one of the known universal properties, it very probably is not a category-theoretical setup. We need a kind of *conceptual categories*, conceptual abstraction instead of structural abstraction. Call such a thing a *conceptopos* or *topocept*?

It is not clear what would be the problematic concept P in general. For the extensions $\mathbb{N} \to \mathbb{Z}, \mathbb{Z} \to \mathbb{Q}, K \to L$ of numbers and fields, one may think of P being an equation, but in the case $\mathbb{Q} \to \mathbb{R}$ we deal with Cauchy sequences, which are not equations in the strict sense. The general situation may not be related to unsolvable equations at all. The first decisive step in this conceptualization is the passage from the naive statement of P to its mathematical shape $S(P)$ as described in Section 10.2.3.5. This procedure does not seem to be a formal/mathematical one, it is the application of the transformed understanding of P in terms of mathematical concepts. Of course, this entire procedure is only valid for type (2) problems. As a such it should also reflect the "problem = solution" approach[6]. How to express this with the extension $V \to V[T]/(P)$?

The problem has two aspects: (1) What is a conceptual extension? (2) What does it mean that the extension is of type (2)?

(1) The general scheme as described in Section 10.2.3.5 first suggests that the conceptual extension must enable to interpret points $x \in V$ as being special instances of the selected concept. For example in the case of the liner equation $a + x = b$, the elements x of $V = \mathbb{N}$ are viewed as solutions of the equation $0 + x = x$. Or for the extension $\mathbb{Q} \to \mathbb{R}$, a rational x is viewed as a constant Cauchy sequence $(x_i = x)_i$. But one and the same $x \in V$ can be viewed as an instance of many concepts, linear equations, polynomials, Cauchy sequences, whatever. There is an entire variaty of concepts that comply with a given domain V. One could also invert the roles and view concepts as points in a concept space C, and for a given $x \in V$ then consider the variety of all concepts that comply with x, i.e., $x^\perp = \{c | c \in C, c(x) = 1\}$, where $c(x) = 1$ means that x can be viewed as an instance of concept c. But we want not one x, we want all of them to be considered, i.e., we focus on the concept set $V^\perp := \bigcap_{x \in V} x^\perp$. Conversely, for $c \in C$, we may define $c^\perp := \{x | c(x) = 1\}$, where the x are taken from a hypothetical domain that comprises V, and for any set $D \subset C$, we define $D^\perp := \bigcap_{c \in D} c^\perp$. We then view the extension of concept c as being the property $V \subset c^\perp$. Conversely, if $c \in V^\perp$, then $V^{\perp\perp} \subset c^\perp$, which implies $V \subset V^{\perp\perp} \subset c^\perp$.—The conceptual extension must be the super space $V \subset c^\perp$ of V. But this space must also inherit the structure of V, this is required from the scheme in Section 10.2.3.5.

---

[6] We should be clear about the proofs of universal properties an their associated objects. These are not, in general, given axiomatically (such as the subobject classifier in a topos), one has to provide a proof. Is this a type (2) situation? I claim that this is indeed the case, although not a trivial one, i.e., it may involve several steps of type (2) problem solving. In view of the general character of a universal property in a category, this however has to be of general abstract character and therefore (!) cannot be of type (3). Type (1) is excluded for obvious reasons.

(2) The second aspect requires that the extension $V \subset c^{\perp}$ comprises the type (2) solution of the problem that is generated by c. This means that we first require that the concept c *is* a problem in V, and second that this problem is solved in the extension. This means that the extension also includes the extension of the structure that is present within V. For example, if $V = \mathbb{N}$, we want the addition and multiplication in $\mathbb{N}$ to be extensible to $c^{\perp}$. This is all but evident in view of the variety of structures that we are facing in the known examples. It is not evident that the conceptual setup of c includes such a structural specification. One could require that extension $V \subset c^{\perp}$ be a structural by definition, but even then, the conceptual setup of c would require to include a structural specification of c, a requirement that is not easy to be made concrete. Even if we interpret instances of c to be instances of object-oriented classes or denotators of given forms, such a structural condition is not evident from the given information.

10.4.1 New Objects Needed?

In view of the fundamental difficulties to understand the ontology of concepts it must be asked whether it is sufficient to work with known mathematical constructions. In Galois' approach, one had to introduces new mathematical structures, fields, groups, etc. Could it be that we now have to introduce concepts as a new type of mathematical objects? Concepts would be this new type. They cannot be described as traditional structures, but must be conceived differently. What then would be a *mathematical concept*? If this works, we must design the behavior of mathematical concepts with the same precision as structures were introduced. Mathematics of concepts must be a new type of mathematics, we must be capable of managing concepts in a mathematical style of thoughts.

A mathematical concept must be capable of comprising mathematical structures, much as a mathematical structure must comprise sets that realize that structure. It is a delicate difference that, for example, a group is not only a concrete set with operations, but the type of operations that are realized in some set. The structure of a group is best described by the category of groups, not by one single instance of the group structure. Perhaps is the "conceptopos" what plays the role of a conceptual 'category': A realm of structures (categories?) that reify a concept. Can we think of mathematical concepts of being instantiated by categories? So what would be the concept of which the category of groups is an instance?

Coming back to the type (2) problems for $\mathbb{N} \to \mathbb{Z}$ and $\mathbb{Z} \to \mathbb{Q}$, these are solutions which only use the structures from the category of commutative and cancelable monoids $(\mathbb{N}, +), (\mathbb{Z}, \cdot)$. But the case of $\mathbb{Q} \to \mathbb{R}$ is a different setup, not in the same category, but still pertaining to the same conceptopos? We have to investigate the difference between conceptopoi and problem types. A specific conceptopos must enable a type (2) problem solution. The idea of a conceptopos remains the main topic now.

So let us look at the situations we know, where type (2) occurs. They are all about conceptopoi that are extensions of given conceptopoi, polynomial algebras, Cauchy sequence spaces, etc. All of them are universal constructions of extensions of conceptopoi collections of categories. This suggests that type (2) problems are solved by canonical solutions already implicit in those universal constructions. This would imply that type (3) problems can be solved by a chain of such universal extensions and their implicit solutions. How could we try to prove such a conjecture, which is, to my mind, the kernel of Grothendieck's philosophy/methodology of mathematics?

A very interesting conceptual extension is the Yoneda construction $\mathcal{C} \to \mathcal{C}^{@}$, which embeds any category into a topos. It is interesting to understand how this extension could be viewed as a problematic one.

So it seems that the general evolution looks as follows:

1. Start on a given body of categories,
2. extend some categories that implement a specific conceptopos by a universal construction,
3. exhibit the new type (2) problems that are generated by this universal construction
4. solve these problems by the innate solution.

This procedure is not automatically the solution of given conjectures, but the Grothendieck methodology states that those conjectures can be solved on the ground of such extensions by a more or less straightforward application of the solutions that are generated by the type (2) problems from these universal extensions.

It could be argued that the general setup of a body of categories is too generic, too simple to enable non-trivial conceptual extensions. This is a wrong argument since the simplicity of such a general setup rightly means that one is free/obliged to add substantial conceptual work. Yoneda, derived categories, etc. are all built upon such simple basic "categorical structure", but they are incredibly efficient. Pierre Cartier's title "Du bon usage d'une tautologie" of his talk at the École normale supérieure in that 2007 conference in Paris about Yoneda's Lemma informs us about the typical misunderstanding of such extensions: they are nearly tautologies, but in their subtlety generate a plethora of non-trivial insights.

The conceptual style of mathematics replaces calculations by thoughts, by conceptual constructions that look trivial from the calculational point of view, but input a conceptual insight that changes the understanding of the actual structures that may have far-reaching consequences for the logic of problem solving. The blind efforts in calculations are replaced by insights into the conceptual architecture.

This means that the conceptual style generates a totally different strategy of gaining insight—replacing blind calculations with hammer and chisel. The Yoneda extension creates the problem of characterizing/describing non-represented functors. The solution, also termed co-Yoneda lemma, characterizes functors as colimits of representable functors over the category of points. This solution has also been coined "near-tautological", because the calculations it includes are trivial. Yes, but the conceptual progress is remarkable. *It is another path to insight, no, it is a path to insight instead of chisel and hammer problem killing.*

A concrete condition of such an extension is that it extends the given domain. For example, the Yoneda embedding is such an extension. But the equation $a^x = b$ in $\mathbb{N}$ isn't an example, because it relates to logarithms, which are not available in that domain. So the embedding condition is the first one. It is however not evident, which concepts might be available.

The next condition requires that the embedding be a structural one, too. This is necessary because the type (2) solution requires that we find objects of the given structure, numbers, that solve the given problems. Therefore one is obliged to extend not only the given domain, but also its structure.

We are now left with the main question: *What are the conditions for a type (2) solution given precisely by the extended 'numbers'?* It is far from evident, why these extended 'numbers' would deliver a type (2) solution. It might be more or less straightforward to take specific quotient polynomial rings, but why the Cauchy sequences would generate the correct new 'numbers', which solve the problem, is not as evident. The question is not a technical one, these configurations are technically trivial. The question is rather *why* such extended 'numbers' are also the solutions of the given problems. The mystery[7] is why these conceptual constructions are also the solutions:

What is the hidden common argument—conceptual logic—behind these constructions?

The naive approach looks as follows in the case of Cauchy sequences: Let the rational Cauchy sequence (a_i) converge to itself as an extended number. This idea must however be verified in two respects: (1) We need a topology on the extended domain. (2) The claimed convergence needs to be verified with this topology. The analogous arguments hold for polynomials: Take the indeterminate T in $K[T]$ as the potential zero of a given polynomial $P(T)$, then create a domain where this is true (passage to the quotient). Let us be clear on the constructive variety that is available. We could also introduce a fictituous point and let every rational Cauchy sequence converge to this point. Of course, this is a brutal idea and also violating many evident requirements. But we have to know what are the many supplementary conditions that we implicitly assume besides the simple convergence. This means that, in the case of Cauchy sequences, we have the specific mystery that declaring (a_i) as the convergence point of the rational Cauchy sequence (a_i) also copes with those supplementary conditions.

[7] It is remarkable that Galois' writings, their style above all, were criticized as being imprecise, what Dieudonné recognized as a style that we nowadays can read as being our contemporary argumentation.

We may define the extended 'numbers' as being the solutions. But then we have to verify that they are as claimed. We already discussed the property of extending the nature of the given numbers, the structural aspect. But now, claiming that we indeed have solutions means to verify that these extended objects solve the given problem.

Summarizing, we are given a domain V. We then have problematic concepts and transform them into mathematically sound objects. We next verify that the objects of V are special instances of the mathematically sound objects, and that these objects also are solutions of the problematic concepts. We next have to verify that these objects are also comprising more general instances (extended 'numbers'). We finally have to verify that these extended objects are also the solutions of the restatement of the original problems in terms of these objects. The last point is the mysterious one: why does this work?

Let us explain this again for the simple extension $\mathbb{N} \to \mathbb{Z}$. We are given the elements $a \in \mathbb{N}$. We want to solve the problematic equation $a + x = b$. To this end we first identify objects a in $\mathbb{N}$ with their defining equation $0 + x = a$. We then encode such an equation by $(0, a)$, or, more general equations $a+x = b$ by the corresponding (a, b). Restating the problematic equation $a+x = b$, we have $(0, a) + x = (0, b)$. The mystery is that the solution is given by $x = (a, b)$, i.e., the encoding of the equation by (a, b) yields the solution. In fact: $(0, a)+(a, b) = (a, a+b) \sim (0, b)$. Here we refer to the equivalence relation $\sim$ among equational encodings, namely $(a, b) \sim (u, v)$ iff $a + v = b + u$. It is remarkable that this solution is completely insensitive to the size of a, b. It works quite generally.

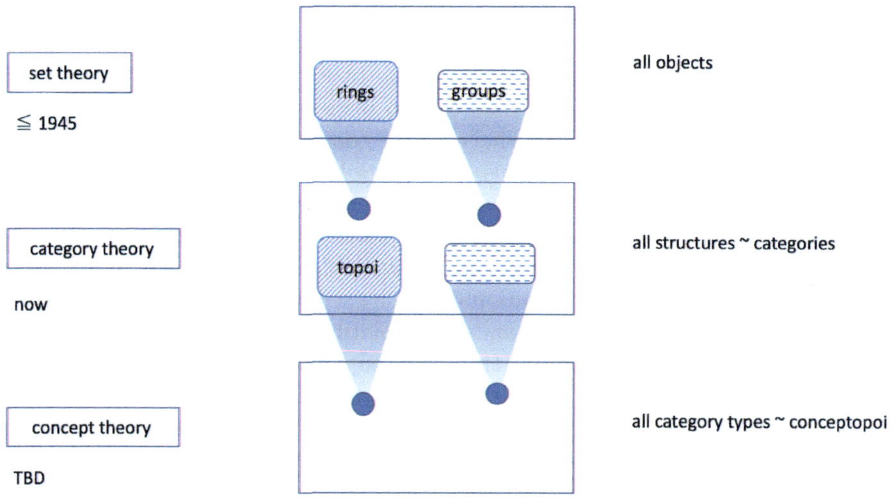

Fig. 10.8: The unfolding of the three mathematical periods: objects, structures, and concepts.

10.4.2 Objects, Structures, Conceptopoi

Mathematics transgressed different stages of theory building, see also Figure 10.8.

The *first stage* focused on single objects, such as natural numbers, real numbers, special function spaces, geometric objects (sphere, tube, Moebius strip, etc.). This was dominant until 1945, when category theory was invented. It had been completed by *set theory*, which was a perfect object factory: you can build every reasonable single mathematical object from set theory.

The *second stage* focused on a description of structures that are imposed upon objects, such as groups, rings, manifolds, etc. The theory that was built around the structural conceptualization was first designed by Bourbaki (starting 1934) and then shaped in an axiomatic style by *category theory* (set up around 1945): A category is the collection of objects of a determined structure, together with a paradigm of structural comparison: morphisms.

The *third stage* (to this date) focuses on conceptual constructions that cover collections of categories. For example, topoi are a conceptual specification of set-like categories. Such collections could be called *conceptopoi*. The very definition and development of this stage is still to be done. It first looks like a conceptopos is defined by structures within categories (such as being cartesian closed, finitely complete and co-complet, having a subobject classifier, etc.).

The difference to structural specifications is however an essential one. A typical structural specification would be to consider categories of commutative groups, rings, skewfields, for example. These are restrictions of given structures to special cases. In contrast, topoi are not only structurally specific, they rather, and importantly so, ask for the addition of specific objects such as subobject classifiers, they add instances of a specific conceptualization. They add concepts, such as also zeroes of polynomials in Galois theory, the extension of fields by new concepts. This looks like a semiotic process: adding (instances of) new concepts to a given semiotic state.

Could this be seen as an evaluation of the (functorial) concept "subobject classifier Ω" at a specific H-jet X: $\Omega_X : X \to \Omega$? An evaluation that typically generates an extension $X \rightarrowtail X^{@}$.

New concepts are not simply there, they have to be added. This extension activity is the difference to structures, which are simply there: a group is a group is a group.

The addition of concepts is a special mathematical style introduced by Galois. Galois theory is not a specific structural theory, it is a conceptual theory. Grothendieck did not understand this either. This step is a deep one, it has been a region of intimacy to mathematicians, they make it, it is essential, but they don't talk about it, they just use it (a new concept) when it's there. It pertains to the *pre-mathematical genesis, it is not mathematical yet.*

This insight is finally the correct point of view!

Pre-mathematics should be transformed into mathematics via the framework of conceptopoi, in the same sense that categories turned pre-mathematical understanding of structures in a mathematical framework.

It is funny how category theory in its first expansion was smiled at as "abstract nonsense" until when Grothendieck taught them the contrary. Yes, but when he resigned, those living deads wanted to go back to the "real, concrete" thing. Well, they tried to, but nowadays, categories are indispensable for all good mathematicians, Grothendieck's achievements, especially topos theory, have created the conceptual machines to move forward. The same will happen to our conceptual mathematics. In 50 years, you will see that this step is also substantial.

The seemingly simple first step is an important wall against creativity, it looks harmless but it encapsulates tons of innovations. Grothendieck's remark about the nearly childish simplicity of the concept of a scheme comes to our mind here. But it revolutionized the entire algebraic geometry. The critical point is that the converse is not true: simplicity does not guarantee depth, it is simply of another nature, simple or not doesn't matter, only the correct conceptualization matters. Period[8].

Kant: concepts are loci/points in a conceptual space. So should we conceive conceptopoi as conceptual spaces which are composed of concept-points? And a conceptopos must be reified in categories like categorical objects can be reified as structured sets. So categories K would be instances of points of conceptopoi, like $K \to J_1, K \to J_2, \ldots$, where J_i are H-jet concepts (in a semantic diagram) of a given conceptopos.

And it is clear now that the understanding of the type (2) problem (problem = solution) must come from conceptual mathematics as it is a conceptual problem regarding the nature of concepts involved in that type. We are still incapable of approaching that situation in a mathematically clean way. Our scheme on page 115 is a pre-mathematical entity. That mystery will be solved as soon as the conceptopological method works.

10.4.3 Concepts and Structures

It seems that conceptopoi have two dimensions: a first dimension $\mathcal{C}$ of proper concepts, and a second one, $\mathcal{S}$, of structures, see Figure 10.9. In this approach, we view the transition $\mathbb{N} \to \mathbb{Z}$ as a passage from $\mathcal{C}$ to $\mathcal{S}$, where the concept C of an equation $a+x=b$ on $\mathbb{N}$ is represented by a set $Str(C)$ of structures, such as $(a,b) \in Str(C)$. In the other direction, a structure S, such as $S=(a,b)$, is represented by a set $Con(S)$ of concepts, which are embodied by S. Define more generally $Str(X) = \bigcup_{C \in X} Str(C)$ for a set $X \subset \mathcal{C}$ of concepts, and also $Con(Y) = \bigcup_{S \in Y} Con(S)$ for a set $Y \subset \mathcal{S}$ of strucures. This entails a topology by the Kuratowski operator $\overline{X} := Con(Str(X))$ (check!). Same on structures:

[8] Mazzola feels radical and deracinated at the same time. We want to get rid of so-called styles of mathematical thinking. Wheelchairs, not thoughts.

$\overline{Y} := Str(Con(Y))$ for a set $Y \subset \mathcal{S}$ of strucures. One may then consider continuous maps bewteen the topological spaces $\mathcal{C}$ and $\mathcal{S}$.

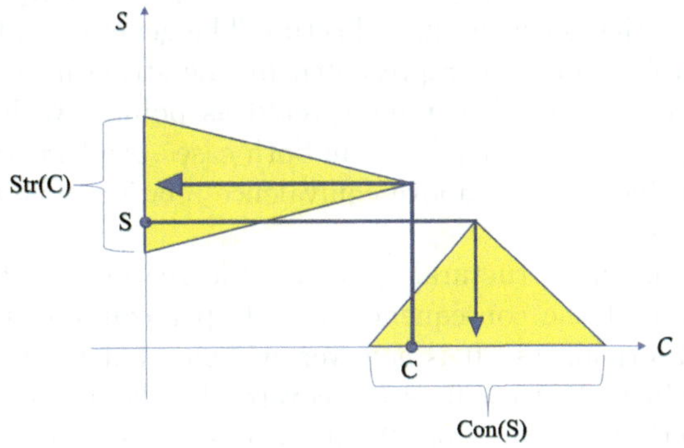

Fig. 10.9: The two-dimensional concept-structure space.

Under this two-dimensional perspective, the movement to solution type (2) becomes quite intriguing: Start with the concept of an equation $C =$"$a+x = b$" on $\mathbb{N}$, map it to $Str(C)$ and exhibit $S = (a, b) \in Str(C)$. Then map this S to the concept C' of a space of points for structure S, to be clear: we reinterpret a structure for an equation as being a point in a space, an extended domain $\mathbb{Z} := \mathbb{N} \times \mathbb{N}/\sim$ (and including the original special set $\mathbb{N}$ of points as pairs $(0, a), a \in \mathbb{N}$). Now, we may introduce the concept C'' of an equation $C =$ "$a + x = b$" on this extended domain of points and find that the generalized point (ex equational structure) (a, b) solves the equation.

10.4.4 A First Synthesis?

It seems that one can unify the two examples: $\mathbb{N} \to \mathbb{Z}$ and $\mathbb{Q} \to \mathbb{R}$ in a shared conceptual setup. In both spaces, $X = \mathbb{N}, \mathbb{Q}$, we have a distance function $d : X \times X \to X$. For $X = \mathbb{N}$, this is the positive difference $d(a, b) = x$ of the equation $a + x = b$ for $a \leq b$ or $b + x = a$ for $b \leq a$. For $X = \mathbb{Q}$, we set $d(a, b) = |a - b|$.

With this, we may look at an equation $a + x = b$ in $\mathbb{N}$ by a generalization. We consider the $\mathbb{Q}$ case: $a_i \to x$ as a case of $a_i + x \to b$, which evidently covers the original Cauchy sequence case. The convergence of a sequence $a_i \to b$ reads as $d(a_i, b)$ becomes arbitrary small for i large. For $\mathbb{N}$, this means $a_i = b$ for i large since distances in $\mathbb{N}$ are natural numbers only. For the $\mathbb{N}$ case, we would take a constant sequence $(a_i = a)$ and then the condition $a_i + x \to b$ reads as $a + x \to b$, which means that (for large i,) we have $a + x = b$, which is the original statement for $\mathbb{N}$.

We also have to deal with the equivalence relation between equations or Cauchy sequences in this unified setup. For the linear equations $a + x =$

$b, u + x = v$ to have the same solution x, the condition is $x = b - a = v - u$, i.e., $b+u = a+v$ in $\mathbb{N}$. For the Cauchy sequence, $a_i+x \to b$ is equivalent to $u_i+x \to v$ iff $b + u_i$ and $a_i + v$ converge to the same number, i.e., $d(b + u_i, a_i + v) \to 0$. This works for $\mathbb{N}$, too: It evidently reduces to the previous equivalence relation.

So we have the following unified setup: The generalized Cauchy sequence $a_i + x \to b$ (in $\mathbb{N}$ or in $\mathbb{Q}$) is represented by the structure $((a_i), b)$, including the structural extension when reinterpreted as points: we have the addition $((a_i), b) + ((u_i), v) = ((a_i + u_i), b + v)$ in both cases. And we see that this operation is compatible with the above equivalence relation. The type (2) solution is also immediate.

Question: Is this structural synthesis the reason for the common type (2) solution or is it the consequence of a deeper commonality beyond similar/assimilated formalisms? It is however evident, that the third example, the polynomial problem $P(X) = 0$, is not covered by the above generalization. It is also not exactly the same type (2) situation since we solve $P(X) = 0$ by the ideal (P) defining the quotient ring, whereas the solution $x = X + (P)$ is not the original polynomial, only nearly so.

10.4.5 Doing Conceptual Mathematics

A major question regarding conceptual mathematics is not only about the mathematical description of mathematical concept architectures; this aspect might be initiated through the H-jet category. It is much more about how to *do* conceptual mathematics. How would "calculation" with concepts look like? Looking back, the structural calculations look very different from the old-fashioned set-theoretical ones: diagram chasing, Yoneda networks, etc.

Our important perspective here is that conceptual mathematics would not be focused on weird calculations, but on conceptual reflections, sometimes trivial from the traditional point of view, but powerful for the solution of a determined problem. Grothendieck's long proofs are a first example of a conceptual procedure as opposed to an intricate "calculation machine".

The example of Section 10.2.3.8 looks like being a first very simple conceptual approach: One wants to generate a domain, where equations $a + x = b$ are always solvable. The idea here is to identify Y (the natural numbers) as a limited concept, limit of the diagram (Y_ι), then simulating the defining projections $Y \to Y_\iota$ for the equational H-jet E, and then looking at the identification of E under these new "projections". This yields the integers as a number domain issued from the conceptual "transport" starting at $Y \to Y_\iota$. Here the effort is not a complicated calculation, but the conceptual transfer operation.

Remark 6. Attention: It is not true that every set-theoretic or category-theoretic problem can be approached easier by conceptual methods. This is already evident for the transition from set theory to category theory. Problems have their individual solution ecosystems. Therefore it is not a priori wise to take any "classical" problem and to try solving it by non-classical methods.

10.4.6 Conceptual Aspects of the Goldbach Conjecture

This section discusses some attempts to manage the famous *Goldbach conjecture* that states that every even natural number ≥ 4 is the sum of two primes. On 7 June 1742, the German mathematician Christian Goldbach wrote a letter to Leonhard Euler (letter XLIII), stating this conjecture. To the date, it has been confirmed up to 4×10^{18}.

Here, we don't want to solve the conjecture but give some ideas about how to address it in a conceptual vein.

To begin with, denote by $Pr[2n]$ the set of odd primes $\leq 2n-3$. To begin with, we want to define a set endomorphism $Q : Pr[2n] \to Pr[2n]$ which is the identity iff its argument q is a Goldbach prime, i.e., iff there is a prime p such that $q+p = 2n$. We may discard n being prime, where Goldbach evidently holds. The first idea behind this approach is to try to invoke the basic theorem of contractions, namely that an endomorphism of a complete metric space that is a contraction has a unique fixed point. Such a fixed point in our context would be a Goldbach prime.

Conceptually speaking, we first have to investigate the difference $2n - q$ for a prime q. So write this difference in terms of prime numbers: $2n - q = p_1^{n_1}.p_2^{n_2}\ldots$, with increasing prime factors $p_1 < p_2 < \ldots$. It happens that $2n - q = p_1^{n_1}, n_1 > 1$, for example $20 - 11 = 3^2$. To take care of the exponent of this single prime, we enrich $Pr[2n]$ adding to each prime a positive natural exponent. In this setup, we denote the elements (q, i) and identify prime q with $(q, 1)$. For the representation $2n-q = p_1^{n_1}.p_2^{n_2}\ldots$, we choose the first prime, together with its exponent, writing (p_1, n_1). So our first map sends q to $P(q) := (p_1, n_1)$. In other words, q is Goldbach, iff $P(q) = (p_1, 1)$ and no other prime $p_i, \ldots, i > 1$ appears in the above factorization. And then $P^2(q, 1) = P(p_1, 1) = (q, 1)$, i.e., the identity on q. If $2n - q = p_1^{n_1}.p_2^{n_2}\ldots$, we define $P(q, m) = (p_1, m.n_1)$. We claim that this map $Q = P^2$ has the required property $Q(q, m) = (q, m)$ iff q is Goldbach. Clearly, if q is Goldbach, the equation holds. Suppose that q is not Goldbach. This means that either $n_1 > 1$ or that $n_1 = 1, n_2 > 0$. In the first case, we have a $Q(q, m) = (p, m.n_1 \ldots)$, and we are done. So let us suppose $2n - q = p_1.p_2^{n_2}\ldots, n_2 > 0$. Then $2n - p_1 = q_1^{m_1}.q_2^{m_2}\ldots$. If $m_1 > 1$, then $P(p_1, m) = (q_1, m.m_1) \neq (q, m)$ and we are done. So suppose $m_1 = 1$: $2n - p_1 = q_1.q_2^{m_2}\ldots$. If $q_1 \neq q$, we are also done, so check $q_1 = q$. So we are left with $2n - p_1 = q.q_2^{m_2}\ldots$, $m_2 > 0$. Multiply this equation by $p_2^{n_2}\ldots$, which means $2n.p_2^{n_2}\ldots - (2n - q) = q.q_2^{m_2}\ldots.p_2^{n_2}\ldots$, i.e., $2n.(p_2^{n_2}\ldots - 1) = q.(q_2^{m_2}\ldots.p_2^{n_2}\ldots - 1)$. This implies $2n | q_2^{m_2}\ldots.p_2^{n_2}\ldots - 1 = \frac{2n-p_1}{q}\frac{2n-q}{p_1} - 1$ and $q | p_2^{n_2}\ldots - 1 = \frac{2n-q}{p_1} - 1$. But if $q > n$, this impossible. This means that for $q > n$, $Q(q, m) = (q, m)$ iff q is Goldbach. $Q(q, m) = (q, m)$ also holds for Goldbach primes $q < n$, but this condition is not sufficient. Since every Goldbach situation in $Pr[2n]$ must involve a prime $q > n$, this manages the problem of fixed points up to the question about a contraction on $Pr[2n]$. Of course, this will not be possible in general since the uniqueness of fixed points will not hold in the Goldbach situation.

Next, we want to give the set $Pr[2n]$ a distance function. Since this set is finite, the corresponding metric space will automatically be complete. We take the distance as being the usual distance $d(p, q) = |p - q|$. This is also the remainder of the division theorem, since all primes are $> n$.

The crucial question now is to find neighborhoods of primes where the Q map is a contraction. Of course, this condition is necessary since the singeton neighborhood of a Goldbach prime $> n$ is such a neighborhood. In other words, if there is no Goldbach prime, this means that every neighborhood is not a contraction. Since the metrical space is discrete, this means that any set of primes $> n$ is not a contraction for Q. Aeqivalently, the distance $d(q, Q(q))$ would be positive for all of $q > n$.

A second, more promising topological interpretation is to refer to the Brouwser fixed point theorem. We recall its special case of a non-existent retraction $F : I \to \{0, 1\} = \partial I$. We may define such a retraction following the classical setup: Take the line $Q(q) \to q$ and the the largest or smallest prime $F(q)$ on the directed line from $Q(q)$ to q within the set of primes $< 2n$ and $> n$, the latter two-element set being the analogue of ∂I. This approach requires a transfer of topological concepts to the present set of primes and the map Q, see also Figure 10.10.

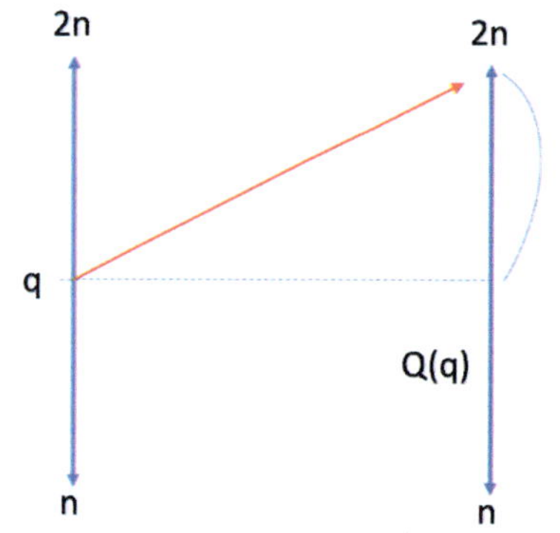

Fig. 10.10: The retraction on primes.

The point of this example is not to solve the Goldbach conjecture (which in itself is not a very attractive statement), but to exemplify the conceptual approach to mathematics. This approach focuses not on calculations or structures, but on the conceptual architecture given by the contraction and fixed point theorems. The former looks quite obvious and not very promising in its local setup, while the second seems more attractive. In the second option, we have to create a map among primes (in the given interval) which simulates the classical retraction, i.e., a map that would not be a retraction if it had a fixed point. And beyond this construction, it would be important conceptually to think about the topological concepts, i.e., topologies and fundamental groups, in the present context. It is however not clear how the topological concept framework might be simulated for the primes' set, instantiating a "discrete" version of Brouwer's fixed point theorem. One approach would consist of the conceptual reshaping of curves without backing topological contents.

Part IV

Applications and Consequences

11

Applications and Consequences

Summary. Music (and more generally the arts) creates semiotic structures independently of 'external' reference contents. We discuss this qualification as a conceptual challenge.

– Σ –

11.1 The Semiotic Power of Music

Summary. We discuss the fascinating fact that music (and more generally the arts) creates semiotic structures independently of 'external' reference contents.

– Σ –

11.1.1 Program and Absolute Music

In musical creativity, we exhibit a number of compositions, which explicitly intend to present semiotic instances, where the musical expression, be it a single sound or a more complex rhythmical, melodic, instrumental, etc., sound construction, points to an extramusical content, such as a thunder, water fall, bird's song, emotional states, etc.

Such a type of music is called "program music", because music refers to an extramusical 'program'. In operas, operettas, musicals, theaters and movies, program music is a well-known phenomenon.

The extramusical contents needn't be existing physical objects or events, they may also be psychological in nature. Music, typically in criminal movies, may point to an emotion, fear, tension, etc. There is a well-defined vocabulary of psychological contents that are incited by musical expressions, such as minor second melodic movements, glissandi of string-like electronic voices, strong rhythmical beats on bass drums, etc. In other words, the program music spectrum extends beyond physical reality contents to psychological significates of all sorts.

G. Mazzola et al., *Functorial Semiotics for Creativity in Music and Mathematics*, Computational Music Science, https://doi.org/10.1007/978-3-030-85190-3_11

When one contrasts "program" music from "absolute" music, in the latter category the above contents are meant to be absent or at least not intended by the music's creator(s). This seems to be a clear dichotomy, but it is much more complicated when one looks at how music is perceived. Music is not simply looked at as an unambiguous objective fact, such as a mathematical formula, this is part of the aesthesic complex.

Music first of all is perceived in its acoustical reality, or also its symbolic surface when given on a score. This expressive starting point is then transformed into a more mental and psychological reality, where the music's content may be built. Music is always interpreted, i.e., transformed into an aesthesic level, a level which is individual, personal, variable, and non-objective.

Within this aesthesic procedure, the intention is always to come up with a semiotic construction, i.e., the construction of semantics from the perceived expressivity. Here again the fundamental semiotic axiom of human behavior is at work. For this reason, aleatoric music cannot be successful since its systematic destruction of structural coherence contradicts human efforts to implement any semiotically valid instances.

So what happens within the aesthesic activity, when an extramusical program cannot be found? Let's say, you are confronted with the last movement of Beethoven's Sonata op. 109. No visible/audible program is at stake, so what? There are two reactions that usually take place: Either you invent an extramusical program or you start being engaged in what is called "absolute" music.

The first alternative is a frequent 'solution' for those who don't display sophisticated analytical tools. They make up an extramusical program, typically in form of a human drama: they set up agents, dramatic roles that are embodied in the music's voices, in the harmonic or rhythmical structures. In such a construction, a tonal modulation may be 'understood' as being a fight between two 'tonal' agents, which eventually is won by the tonality agent where the music moves. This imposed extramusical semantics does not solve the music's structures per se, but gives them a psychological force field that simulates what the music is performing. This strategy, unfortunately quite frequent even in academic contexts [6], replaces the structural understanding by an extramusical simulation that is based upon insinuated programmatic pointers.

11.1.2 The Reference Architecture and Consistency in Absolute Music

After having dealt with programmatic realities and make-ups in music, we are left with the question of how the semiotic architecture of "absolute" music may be constructed. We suppose here that a musical semiotics is being built that does not refer to extramusical contents, but nevertheless displays a consistent semiotic structure. This situation is similar to the mathematical reality, where you cannot refer to physics or psychology, when dealing with mathematical contents.

Such a consistent semiotic architecture needs to rely on two fundamental pillars:

- a *conceptual setup* whose H-jets refer to each other and nothing else, in music theory this may include the concepts of notes, with their parameters (onset, pitch, loudness, duration, voice, glissando, crescendo), groups of notes such as chords, melodies, rhythms, groups of groups such as cadences, tonalities, etc., together with the mathematical and/or physical auxiliary concepts;
- a logically consistent discoursive scheme to deduce statements from given statements or axioms, a *musical logic.*

Using this semiotic methodology, one may build a musical composition's immanent semiotic skeleton. See our examples in Sections 4.1, 4.4, 4.6.4, 10.2.1.4. It is essential for this "absolute" approach that it doesn't refer to extramusical contents of psychological, sociological or physical ontology. The latter means that one may refer to a sound's frequency, but not to a gun or other semiotically loaded objects, or if such objets are needed, their semiotic content would not be used.

The question of the content of a musical composition/improvisation in terms of beauty, of aesthetic quality, is referring to beauty in abstract structures. It includes the concept of beauty in mathematical or geometric entities. For example the beauty of a icosahedron, or the beauty of a gestural analysis of Beethoven's initial fanfare [28, Section 62.4.3]. This type of content is radically different from psychological perspectives. It also includes Jakobson's poetic function, which is a classical semiotic shape beyond psychology.

11.2 Implementation Issues

Summary. We discuss the option of implementing semiotic structures and its problems.

$- \Sigma -$

11.2.1 Object-Oriented Implementation in Java

Summary. Implementing semiotic structures in the object-oriented paradigm in Java language.

$- \Sigma -$

In this section we want to sketch a Java-programmed architecture of the categories that are derived from the semiotic conceptualization of H-jets and their relationships.

To begin with, the basic class should be the sign structure, comprising expression, signification, and content. This would look a follows:

```
class public Sign{
     public Sign expression;
     public Sign signification;
     public Sign content;
     }
```

For specific sign properties, one has to defined subclasses of Sign, such as, for example,

```
class public NamedSign extends Sign{
     public String SignName;
     }
```

Since with this construction, every sign refers to its three components, we also need to include the case, where a sign is initial, i.e., does not refer to such defining components. This is dealt with by setting its instantiation with three "self" fields.

For the definition of a H-jet, we need to specify a list of references "from bottom", starting with a bottom sign and then referring to its component, and from this component to its component, etc.

This is done by the class

```
class public HJet{
     public Sign bottom;
     public Int[] references;
     }
```

The list `references` is the list of references of signs to one of their components. For example, the Int array list $1, 2, 0, 0$ would first refer to the signification of the `bottom` sign, then to the content of this signification sign, then to the expression of this content sign, and again to the expression of this expression.

To define morphisms of H-jets, we first need the case of elementary "paths". This is defined by the class

```
class public MorphElement{
     public HJet domain;
     public HJet codomain;
     public Int domainReference;
     public Int codomainReference;
     public Sign Intersection;
     }
```

Finally, to describe general morphisms, which are the paths composed of elementary paths, we define

```
class public Morphism{
     public MorphElement[] ElementSequence;
     }
```

Of course, we need a number of important methods, e.g., to decide upon the compatibility of the intersection for a MorphElement object. Or the length of a H-jet list, or the composition of morphisms, special constructors, etc.

It will be useful to implement a useful set of such classes to get off the ground. And it would be also useful to implement a rubette network in the Rubato environment to manipulate situations in the H-jet categories, especially for the definition of new sign classes. The temporal growth of the category of H-jets is essential here.

Part V

Conclusions

12

Conclusions and Perspectives

Summary. This chapter reviews how the mathematics of new concepts of music that we discuss in the book can be used to better understand creativity and its logic. The reading of this book will support your ability to become a more versatile music creator. By tying together the concepts of functorial semiotics, you will expand your ability to compose methodically and strategically, rather than simply waiting for inspiration to strike.

– Σ –

12.1 Intelligence and Creativity Advancement with Mathematics

In the book *Cool Math for Hot Music*, Mazzola et al. proved that innovative mathematical re-conceptualization leads to explicit and efficient modes of creative tonal modulation. [33, Ch. 23.4]. The new conceptualization was realized in a geometric setup, which enabled the composer to shape modulation in a given composition using a toolbox of geometric objects.

12.1.1 Physics

We could compare the role of mathematics in music to its role in physics. Calculus supports physical formulas, so having mathematics as a new language made the development of physics possible. Mathematical language modeling, a framework derived from calculus, was a new language that described physical movement. Thus the extension of language is an essential step in understanding physics. Similarly, understanding the creative kernel of music needs mathematical conceptualization. Its substance is in constant expansion. Mathematics lends a supportive language tool to physics as it does to music.

G. Mazzola et al., *Functorial Semiotics for Creativity in Music and Mathematics*, Computational Music Science, https://doi.org/10.1007/978-3-030-85190-3_12

12.1.2 Linguistics

Language is like the current of a river, ever-changing. Why not use semiotics of language to support music creation? The German idiom "etwas zur Sprache bringen" (to carry something to the language) – this never-changing language idea must be criticized. The language must be adopted and "moved" to the given problem. This semiotics procedure becomes evident in a thorough discussion of the process of musical and mathematical creativity. In other words, understanding creativity and the option of giving it conceptual drive heavily depends on the question of how new signs and contents could be conceived in the framework of computational formalism.

12.1.3 Quantum Mechanics

It's the modern language of Hilbert space linear algebra that describes the reality of Quantum Mechanics. In this language, a physical "observable" is a linear endomorphism of the Hilbert space, and its measurable projection in physical reality is given by its eigenvalues. Therefore, the physical reality is split into an abstract Hilbert space on the one hand, and the observables' eigenvalues on the other. This is a radically different way of describing physical reality. For more details, check [29, Ch. K.4].

12.1.4 Music

Free jazz pianist Cecil Taylor describes his fundamental ideas as a search for a new language. This new language is a gestural approach as described in Section 10.2.1.2.

The conceptual direction has one significant deficiency: it emphasizes the pure form of the mathematical approach, which fosters abstraction from semiotic aspects of these phenomena. Focusing on forms reduces the adequacy of one's attention: the contents vanish.

12.1.4.1 The Role and Importance of Semiotics in Musical Creativity

Reading composition is communicating via semiotics (e.g., reading scores). The nature of the composition means to recombine existing signs or to add a new sign system. For example, Arnold Schoenberg used the existing notation system and created his 12-tone musical compositions. On the other hand, Hans-Christoph Steiner's score for Solitude was created using *Pure Data's* [38] forms (see Figure 12.1).

Creating new signs is the creation of more nuanced ways of communication. Creation involves re-arranging signs or introducing a nuanced addition of new signs and extending a given semiotic system.

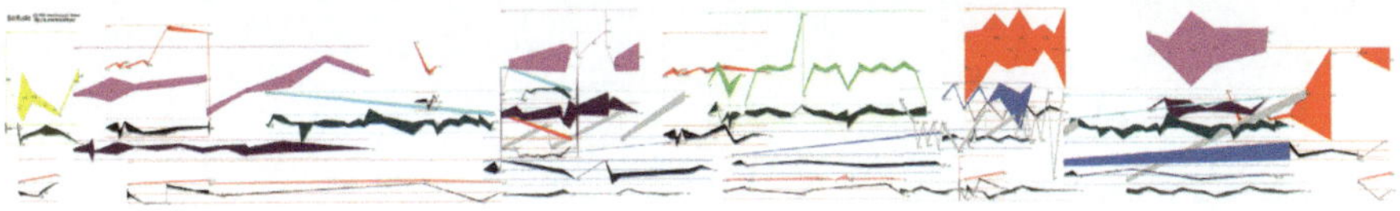

Fig. 12.1: Hans-Christoph Steiner's score for Solitude.

12.1.4.2 Psychological Aspects of Semiotic Activity for Creativity

Creativity is practical without semiotic extensions, however, we argue that the semiotic system needs extensions to discover and push for new possibilities. For example, in Chapter 2.4, *The Babushka Principle in Semiotics: Connection, Motivation, and Metatheory*, we have discussed that the process of condensing and expanding sign systems is responsible for the translation of the score into a performance, and ultimately to the listener's interpretation of the performance. One can even argue that sign systems underlie the way in which society views a piece of music or a certain performer. Such an assertion is reminiscent of the psychological concept of schema formation, in which individuals form an understanding of something new through activating an intricate, underlying web of connections of related concepts. This expressive starting point is then transformed into a more mental and psychological reality, where the music's content may be built. Music is always interpreted, i.e., transformed into an aesthetic and aesthesic level, a level that is individual, personal, variable, and non-objective (refer to Chapter 11).

"Program music" refers to an extramusical 'program'. The extramusical contents needn't be existing physical objects or events, they may also be psychological in nature. When an extramusical program cannot be found, there are two reactions that usually take place among listeners: either you invent an extramusical program or you start being engaged in what is called "absolute" music. This should be a semiotic endeavor, where you don't stop searching for the "musical idea" of your engagement.

Impact. This book deals with musical creativity by applying a mathematical foundation of semiotics to its role in human creativity in general. We have investigated the question concerning a formal, or rather, mathematically precise, representation of semiotics.

This book opens up new questions about the relationship between mathematics and semiotics, and it introduces composers to the possibilities and challenges of creating a comprehensive and computational semiotic theory of creativity in music. Precise analysis of creative processes is the the most important topic in the present world. Our book advances especially the discussion of the creative process in music.

– Θ –

Part VI

References, Index

References

1. Abel N H: Oeuvres complètes, L. Sylow and S. Lie (eds.). 2 vols, vol. II, Christiania 1881
2. Agustín-Aquino O A, J Junod, G Mazzola: Computational Counterpoint Worlds. Springer Series Computational Music Science, Heidelberg 2015
3. Alexander S: Jazz of Physics. Basic Books 2016
4. Appel K and Haken W: Every planar map is four colorable. Illinois J of Math., vol. 21, pp. 429-567, 1977
5. Barthes R: Éléments de sémiologie. Communications 4/1964
6. Cherlin M: Schoenberg's Musical Imagination. Cambridge University Press 2007
7. Coffmann D: Effects of Mental Practice, Physical Practice, and Knowledge of Results on Piano Performance. Journal of Research in Music Education, 38(3), pp. 187-196, 1990
8. Cybenko G: Complexity Theory of Neural Networks and Classification Problems. In: Almeida L B, C J Wellekens (Eds.): Neural Networks. Springer LN in Comp. Sci. 412, 1990
9. Dowek G: Les Métamorphoses du calcul. Le Pommier, Paris 2007
10. Finsler P: Aufsätze zur Mengenlehre. Unger G (ed.), Wiss. Buchgesellschaft, Darmstadt 1975
11. Gabriel P and M Zisman: Calculus of Fractions and Homotopy Theory. Springer, Heidelberg 1967
12. Fillmore, Ch J: Frame semantics and the nature of language. Annals of the New York Academy of Sciences, vol 280, pp. 20-32, 1976
13. Goguen J: An Introduction to Algebraic Semiotics, with Application to User Interface Design. Springer, Heidelberg 1999
14. Grothendieck A, J A Dieudonné: Éléments de Géométrie Algébrique I. Springer, Heidelberg 1971
15. Hjelmslev L: Prolégomènes à une théorie du langage. Minuit, Paris 1968-71
16. Hellegouarch Y: Invitation to the Mathematics of Fermat-Wiles. Academic Press, San Diego 2002
17. Jakobson R: Linguistics and Poetics. In: Seboek, TA (ed.): Style in Language. Wiley, New York 1960
18. Lawvere F W and S H Schanuel: Conceptual Mathematics. ambridge U Press 1997

G. Mazzola et al., *Functorial Semiotics for Creativity in Music and Mathematics*, Computational Music Science, https://doi.org/10.1007/978-3-030-85190-3

19. Lawvere F W: The Category of Categories as a Foundation for Mathematics. Conference on Categorical Algebra 1965
20. Lawvere F W: Functorial Semantics of Algebraic Theories and Some Algebraic Problems in the Context of Functorial SemanticsS of Algebraic Theories. Reprints in Theory and Applications of Categories, No. 5, pp. 1-121 2004
21. Kendon A: Gesture: Visible Action as Utterance. Cambridge University Press, Cambridge 2004
22. Manin Y: We Do Not Choose Mathematics as Our Profession, It Chooses Us. 1268 NOTICES OF THE AMS VOL. 56, No. 10, 2009
23. Mazzola G: Gruppen und Kategorien in der Musik. Heldermann, Berlin 1985
24. Mazzola G and Hofmann G R: Der Music Designer MD-Z71 - Hardware und Software für die Mathematische Musiktheorie. In: Petsche H (ed.): Musik - Gehirn - Spiel, Beiträge zum 4. Herbert-von-Karajan-Symposion. Birkhäuser, Basel 1989
25. Mazzola G: *Synthesis*. SToA 1001.90, Zürich 1990
26. Mazzola G: Semiotic Aspects of Musicology: Semiotics of Music. In: Posner R et al. (Eds.): A Handbook on the Sign-Theoretic Foundations of Nature and Culture. Walter de Gruyter, Berlin and New York 1998
27. Mazzola G et al.: The Topos of Music—Geometric Logic of Concepts, Theory, and Performance. Birkhäuser, Basel et al. 2002
28. Mazzola G et al.: The Topos of Music III: Gestures. (Third volume of second edition of [?]) Springer, Heidelberg 2018
29. Mazzola G et al.: The Topos of Music IV: Roots. (Fourth volume of second edition of [?]) Springer, Heidelberg 2018
30. Mazzola G, G Milmeister, J Weissmann: Comprehensive Mathematics for Computer Scientists, Vols. I, II. Springer, Heidelberg et al. 2004
31. Mazzola G: Categorical Gestures, the Diamond Conjecture, Lewin's Question, and the Hammerklavier Sonata. Journal of Mathematics and Music Vol. 3, no. 1, 2009
32. Mazzola G, J Park, F Thalmann: Musical Creativity. Springer, Heidelberg 2011
33. Mazzola G, M Mannone, Y Pang: Cool Math for Hot Music. Springer, Heidelberg 2016
34. Mazzola G, M Mannone, M O'Brien, Y Pang, N Torunsky: All About Music. Springer, Heidelberg 2016
35. Mazzola G: Functorial Semiotics for Creativity. Journal of Mathematics and Music 2020, Vol. 14, no. 1, 2019
36. Mac Lane S, I Moerdijk: Sheaves in Geometry and Logic. Springer Heidelberg 1992
social and cultural context (pp. 33–56). Gainesville: University Presses of Florida, 1981
production-systems/maschine
37. Popper K and J Eccles: The Self and Its Brain. Routledge 2014
38. `https://puredata.info`
39. Saussure F de: Cours de Linguistique Générale. Payot, Paris 1922
40. Sawyer R K: Group Creativity: Music, Theater, Collaboration. Lawrence Erlbaum Associates, Mahwah 2003
41. Tarasti E (Ed.): Musical Signification, Essays in the Semiotic Theory and Analysis of Music. W de Gruyter pp/189-196, 1995
42. Tomasello M: Why Don't Apes Point? In Enfield N J & S C Levinson (Eds.). Roots of Human Sociality: Culture, Cognition and Interaction, Berg, Oxford & New York, pp. 506-524, 1996

43. Tomasello M et al.: A New Look at Infant Pointing. Child Development Vol. 78, No. 3, pp. 705-722, 2007
44. ZIF Conference: From Computational Creativity to Creativity Science. University of Bielefeld, Center for Interdisciplinary Research, Se[t. 2016

Index

G. Mazzola et al., *Functorial Semiotics for Creativity in Music and Mathematics*, Computational Music Science, https://doi.org/10.1007/978-3-030-85190-3

9783030851897